Body Integrity Dysphoria and the Ethical Dilemma of On-Demand Amputation

Leandro Loriga

Body Integrity Dysphoria and the Ethical Dilemma of On-Demand Amputation

Redefining Wholeness: Identity, Autonomy, and the Moral Boundaries of the Human Body

Springer

Leandro Loriga
Department of Medical Ethics, Faculty of Medicine
Masaryk University
Brno, Czech Republic

ISBN 978-3-031-82442-5 ISBN 978-3-031-82443-2 (eBook)
https://doi.org/10.1007/978-3-031-82443-2

This Springer imprint is published by the registered company Springer Nature Switzerland AG
The registered company address is: Gewerbestrasse 11, 6330 Cham, Switzerland

To Jana, thank you for keeping me afloat, always encouraging me to reach my full potential, and supporting every step of my many endeavours with unwavering patience.

Preface

This book addresses the issue of Body Integrity Dysphoria (BID) and its implications across several interconnected fields: medicine, bioethics and medical anthropology. The goal is to provide an exploration of BID, with a focus on the ethical, philosophical and sociocultural dimensions that shape how this condition is understood by individuals who have made the medical practice their profession.

I hope this work will serve as a valuable aid for all those interested in the ethical and practical challenges of on-demand amputation by offering a new perspective on the matter.

The idea that otherwise mentally and physically healthy individuals want to have a part of their body surgically removed is provocative, and challenges cultural perceptions of what does it mean to be *whole* or *healthy*.

This research is inherently shaped by my background in medical anthropology and bioethics which has influenced both the theoretical framework and the empirical methodologies employed throughout this work. This interdisciplinary approach integrates anthropological insights with bioethical analysis, resulting in a multifaced exploration. My nearly decade-long engagement with BID, which continues to this day, has offered me the opportunity to interact closely with healthcare professionals as well as individuals affected by this condition. However, this work specifically focuses on the perspectives of medical professionals and does not include the experiences of those with BID or those claiming to have it. Additionally, although I have encountered through the years black-market type of dynamics to fulfil the amputation desire, this aspect remains beyond the scope of this work. As a culmination of my PhD research, this book marks a significant milestone in this ongoing journey. It combines theoretical insights with extensive empirical work for what will be, I hope, a new understanding of BID and its ethical implications.

Drawing on 36 months of extensive fieldwork and a nationwide survey of over a thousand physicians, this research examines the bioethical principle of autonomy from the perspectives of various medical professionals, particularly those in Cosmetic and Reconstructive Surgery.

One of the key findings, and a provocative claim of this work, is that requests for on-demand amputations, such as those associated with BID, are perceived as anomalies that disqualify individuals from being regarded as fully autonomous agents. This disqualification highlights a lack of logical consistency in the autonomy assessment procedures conducted by the medical community, which are frequently shaped by subjective and cultural biases regarding the sanctity and ideal natural form of the human body. As a result, individuals who successfully navigate the rigorous loops of medical assessment and are recognised as fully autonomous agents capable of making informed medical decisions are suddenly disqualified from this status the moment they express such apparently unconventional requests. This supports the idea that autonomy is viewed as an inherent characteristic of the individual up to a certain point beyond which it becomes a characteristic of the actions or intentions expressed. In other words, certain intentions can invalidate a previously established assessment of autonomy, no matter how well-founded it was before the intention was expressed. When this happens, the individual loses their status as a peer to the physician, allowing the latter to take full control of the medical assessment, thereby reinforcing an outdated notion of medical paternalism.

BID is now recognised as a medical condition, yet it lacks an established therapeutic pathway. Such an uncertainty leaves individuals with BID in a precarious position, where their autonomy and requests are often outright dismissed without proper and thoughtful assessment.

Although this research is rooted in the context of the Czech Republic, its significance extends far beyond national borders. The findings and insights presented in this book are intended to contribute to the global discourse on Body Integrity Dysphoria, engaging with and enriching contemporary literature on bioethics, medical autonomy, and the social understanding of the body. By focusing on universal themes and excluding region-specific bioethical debates, this work aims to resonate with a broad international audience. The issues explored here are not confined to any single cultural or medical context; they are deeply relevant to ongoing global conversations about the nature of autonomy, identity and the ethics of medical decision-making. Through this lens, the book seeks to offer perspectives that are both timely and universally applicable, engaging with current academic debates and challenging established norms in the broader field of bioethics.

Brno, Czech Republic Leandro Loriga

Acknowledgements

I am deeply grateful to the individuals whose support and guidance have been instrumental throughout this journey.

First and foremost, I extend my heartfelt thanks to Prof. Josef Kuře, head of the Department of Medical Ethics, Faculty of Medicine, at Masaryk University in Brno, whose belief in this project allowed me to pursue research under his supervision. His support not only shaped the direction of this work but also connected me with scholars whose insights were invaluable to its development. His influence has profoundly impacted my academic journey and life choices.

I am also immensely grateful to Prof. Iva Šmídová and Prof. Radim Bělohrad for their supervision over the years. Their mentorship has fundamentally shaped my intellectual approach, broadened my horizons and deepened my understanding of complex issues. Their guidance will continue to resonate with me throughout my career.

To Prof. Eliška Bartečková, I offer my sincere thanks for always being available and creating opportunities for rich and meaningful interactions. Her expertise provided critical insights that significantly enriched my research, and her support has been deeply appreciated.

I also wish to thank Prof. Roberto Brigati from the University of Bologna, who ignited my passion for research and offered steadfast support. His influence extended beyond professional guidance, fostering a cherished friendship that has been a cornerstone of this journey.

Finally, I extend my gratitude to Masaryk University in Brno, an institution of excellence where I had the privilege to work and study. Although I cannot personally thank every individual, I appreciate the exceptional environment, opportunities and the incredible people who make this university a remarkable place for academic and personal growth.

Contents

1 Identity, Gender and Narrative in BID 1

2 Theoretical Framework: Understanding BID Through Embodiment 11

3 The Ethics of Body-Modifying Surgeries 25

4 Legitimacy of Bodily Modification Intervenstions 37

5 Pre-operative Risk Assessment and Post-operative Risk Behaviours 49

6 Moral Responsibility 59

7 Blame and Amputation: Uncovering Preconceptions and Addressing Physicians' Scepticism 69

8 Health and Well-Being: Philosophical Approaches 81

9 Closing Remarks: Voluntary Limb Removal and the Challenge to Patient Autonomy 109

Index 125

Chapter 1
Identity, Gender and Narrative in BID

1.1 Introduction

Body Integrity Dysphoria (BID) is a complex condition characterised by an intense and persistent desire to become physically impaired. According to the ICD-11, this desire can lead to self-mutilation, posing dilemmas for both patients and healthcare professionals regarding treatment options. Over the years, symptoms comparable to those of BID have been classified under various names, such as Body Identity Integrity Disorder, Body Dysmorphic Disorder, Xenomelia and Amputee Identity Disorder. However, it was only with the ICD-11 that BID received an official definition, legitimising it within the medical framework, although without establishing any official therapeutic approach.

BID has long been subjected to public attention, often mischaracterised as a mental disorder or paraphilia, resulting in significant stigmatisation for those affected. Despite this attention, there has been a notable lack of literature. Most existing studies focus primarily on its neurological aspects, often neglecting the social and ethical dimensions of the condition.

Amputation in the context of BID represents a significant departure from traditional norms and values associated with the body. It can be seen as a symbolic rejection of conventionality and an embrace of individualism and self-expression, challenging deeply ingrained cultural and societal norms about the sanctity of the human form. In this sense, it embodies a rebellion against the constraints of Modernism, favouring Postmodernism's rejection of grand narratives and its embrace of fluidity, fragmentation and subjectivity.

L. Loriga, *Body Integrity Dysphoria and the Ethical Dilemma of On-Demand Amputation*, https://doi.org/10.1007/978-3-031-82443-2_1

1.2 Autonomy

When talking about bioethics, autonomy often refers to self-determination and self-governance and entails normative and practical characteristics that the exercising entity has to possess. Self-determination and self-governance do not imply that autonomy is solely a negative right, namely the exercising entity's right to be free from external interferences, but also the positive right to seek and obtain the needed information to make an autonomous choice, as it might be the case for the informed consent process (Wilson, 2007, p. 354). The Western conception of autonomy highlights particularly its positive aspect (Beauchamp & Childress, 2013, pp. 101–141; Berlin, 2002, pp. 178–181; Veatch & Guidry-Grimes, 2020, pp. 133–137), which is the duty of the medical world (but also of the broader sociocultural context) to provide the necessary support to have a situational and consequential understanding of the desired action. The underlying idea is that the subject often cannot make a meaningful decision without assistance from others.

From a healthcare perspective, autonomy becomes a particularly complex concept when individuals request invasive medical procedures despite showing no visible signs of physical suffering. For instance, elective surgeries[1] can lead to prolonged physical and psychological distress, raising significant ethical concerns. In these circumstances, the patient's right to self-determination and self-governance can contradict the physician's principle of non-maleficence. The patient has no right to demand any procedure not in the interest of their overall health.[2] In this regard, the concept of autonomy is limited as it would result in a negative outcome of a patient-specific risk-benefit analysis (Crerand et al., 2008, pp. 7–8; Nejadsarvari et al., 2016).

This approach brings the idea of authenticity to the foreground, which plays a fundamental part in ontology and epistemology, and has a normative and evaluative power. To be recognised as authentic (this is to say, to have a condition medically recognised) means to have access to a series of medical diagnoses and interventions otherwise unreachable for the individual. Conversely, to be recognised as inauthentic (that is, proposing a condition which is not present in any of the medical taxonomies used nowadays) leads to being less autonomous—from a normative perspective—or not autonomous at all. Another fundamental aspect is the individual's decision-making capacity, which is the patient's ability to understand the risk-benefit ratio of a given procedure and be aware of any possible alternative to the proposed treatment or intervention. In this case, an individual's decision-making capacity, or competency, is determined by the specific task at hand. This is what Buchanan and Brock call competency determination, which is "a determination of a particular person's capacity to perform a particular decision-making task at a

[1] The term *elective* is vague and subject-dependant. I will clarify its meaning in the following chapters.

[2] The question of who holds the authority to determine what is best for an individual's health is a complex issue that will be explored in detail in the following pages.

particular time and under specific conditions" (Buchanan & Brock, 1986, p. 23). Such definition entails the agent's malleability to reframe priorities and desires according to the newly provided information.

One central question will guide this book (see, for example also, Bayne & Levy, 2005): under what conditions can it be considered reasonable for doctors to agree to on-demand amputation requests? Such central questions will be addressed by implementing three parallel but complementary investigations:

- How is the biomedical principle of autonomy interpreted and assessed from the medical standpoint regarding the aetiology and treatment of elective medical bodily modifications?
- To what extent does a patient's autonomy conflict with the physician's obligation of beneficence and non-maleficence?
- How do physicians discriminate between legitimate and illegitimate requests of bodily modifying intervention?

In other words, the task at hand is to assess what criteria the medical field uses today to approve or deny medical procedures that are based solely on the patient's desires and not on any objective physiological evidence. To answer such questions, one needs to investigate how the biomedical principle of autonomy is interpreted and assessed from the medical standpoint regarding the aetiology and treatment of elective medical bodily modifications. It is essential to inquire about what factors the medical world considers when evaluating this principle and how it discerns such rightfulness of purpose from the requesting entity. Through an analysis and comparison of surgeries due to Gender Dysphoria, including sex reassignment surgery (or gender-affirming surgery) and plastic and aesthetic surgeries, the BID phenomenon can be used as a reference point to inquire into the discriminating factors that lead the medical personnel to act or refrain from performing bodily modifying interventions by providing a critical ethical analysis.

1.3 Identity

The concept of identity has many declinations. Brubaker and Cooper note that there are two different ways to approach the identity problem and that none of these approaches may be considered appropriate for its analysis. The first of these approaches conceive identity as a category of practice—operational—where the term identity is used mainly by social actors in everyday life for political claims. The second approach conceives identity as an analytical category and is primarily employed in the social sciences.

> By "categories of practice", following Bourdieu, we mean something akin to what others have called "native" or "folk" or "lay" categories. These are categories of everyday social experience developed and deployed by ordinary social actors, distinguished from the experience-distant categories used by social analysts. (Brubaker & Cooper, 2000, p. 4)

Suppose a movement against discrimination rooted in what is believed to be identity-based terminology such as African, Hispanic, Disabled and others. In such a situation, there would be both the experience of discrimination faced by people (as was the case with Rosa Parks for the Montgomery Bus Boycott) and the subsequent theoretical elaboration. In this latter variation, the concept of identity undergoes revisions in more fluid and constructivist terms, making it more flexible than a category of practice. This led to a proliferation of meanings so broad as to make the term almost unusable for scientific purposes. For the social sciences, identity is seen more as an artificial product than a natural occurrence. However, things are different for the social actors who conceive the discourse on identity not as the product of a construction and negotiation process but as a natural manifestation of their being. Identity reveals a consensus regarding human nature and a statement of historical and political presence, as might be true for ethnic or religious minorities (Leve, 2011). In this latter sense, identity is a powerful organisational tool in everyday life that reflects the need for social action. One putative issue with such a vision is that identity is conceived as objective rather than malleable. Such a short elaboration on the potential issues that one might encounter when working with the concept of identity was necessary to clear the path for further explanation and lay down the basis for the relevance of a coherent life narrative for BID individuals.

1.4 Gender and Narrative

For the current purpose, the concept of identity will be tacked both from an analytical and an operational perspective, with the latter intended as the feeling of being constant over time and a definite entity. It is built and based on the subject's reflections when comparing oneself with others. George Herbert Mead, for example, roots the idea of identity in sociality, which is "the capacity of being several things at once" (Mead, 2002, p. 49); in other words, covering different social roles throughout time and space without losing one's sense of Self. Such capacity entails, among others, taking others' perspectives to reflect upon oneself (Mead, 2009, pp. 152–164).

As might be the case, however, for specific instances, the identity of the individual does not necessarily reflect one's physical appearance, creating an incongruence that raises a sense of dysphoria (or one's sense of dysphoria may lead one to seek other types of identities). One similar situation liable to be exploited to expand on the dysphoria phenomenon experienced by BID individuals is Gender Incongruence or Gender Dysphoria (depending on the medical taxonomy used). One specific aspect of personal identity construction is the genesis of gender identity, which does not inevitably overlap with sexual characteristics. If sex differences are biological, gender differences are due to psychological, social and cultural factors.

In past times, no differences were speculated between the biological sex of the individual and one's sex typing, which was seen as the natural adaptive response to the surrounding sociocultural context. Cross-sex typing, on the other hand—namely, the incongruence with the surrounding sociocultural gender norms—was deemed deviant from the norm and harmful (Kagan, 1974, pp. 137–167). It was only later that such vision was challenged by a more malleable approach to cross-sex typing, which interpreted such variation not as a deviation but as subjective positive adjustments (Bem, 1974, 1981). Contemporary approaches to gender theories speculate a multifactorial influence on one's gender development (David et al., 2013, pp. 139–140; Powlishta, 2013, pp. 107–108; Ruble et al., 2013, pp. 38–39) with a particular focus on a three-dimensional approach focusing on the centrality of gender in one's identity, evaluation of gender-related social values as well as social pressure to adapt to traditional gender roles (Lurye et al., 2008).

It is necessary now to articulate why it was deemed indispensable to introduce such apparent diversion concerning gender before retrieving and moving forward on the concept of narrative identity and its relevance to BID. Such relevance comes from three points, namely the long-lasting theoretical elaboration that led gender to be seen as something malleable (namely Cross-sex typing) rather than something bound to natural and unavoidable occurrences, how such long-lasting elaboration influenced the present medical categories and revolutionised the way sex and gender are seen not only at the social level but also from a medical perspective (the depathologisation of homosexuality as an example) and finally the possibility to envision and accept the possibility of an individual experiencing a state of dysphoria caused by the incongruence between one's Self and one's physical appearance as it might be the case for Gender Dysphoria.

Similarly to Gender Dysphoria, BID, as of today, sits on the same premises, which is the incongruence between one's Self and one's physical appearance. Contrary to Gender Dysphoria, however, the BID phenomenon does not yet have the same tradition of long-lasting theoretical elaboration to account for it. Similarly, no medical consensus exists on the condition, both in terms of aetiology and treatment. As a further point, it is necessary to mention that for Gender Incongruence (or Gender Dysphoria) and BID, narrative, in both cases, can play a predominant role in helping the individual make sense of oneself and in helping the surrounding environment make sense of the individual. The following section will contextualise the concept of narrative identity as applied to medical cases.

1.5 Narrative Identity as Symptomatology

The growing interest in narratives can be traced back in part to the impact of postmodernism with its rejections of meta-narratives—"large-scale theoretical interpretations purportedly of universal application" (Harvey, 1989, p. 8)—and their totalising character (Harvey, 1989, pp. 39–45), which calls into question the idea of objective truth, pointing out how this can be provisional and contingent on the context. The patient's narrative is reinterpreted with new cognitive and emotional meanings to help them negotiate a social identity affected by their illness and to give moral significance to their actions and experiences. This narrative becomes especially important in cases where symptoms lack a clear physical cause, such as with some psychiatric disorders or general states of dysphoria. In these circumstances, the narrative can be seen as a means of rehabilitation and recovery, especially for those disorders that can be traced back to a sense of self-fragmentation (Robert, 1999, pp. 155–156).

> When people experience illnesses and health conditions requiring considerable medical intervention, they monitor and discuss their bodies. How they talk about their bodies tells us much about the nature of embodiment and about how gender, age, and ethnicity influence the way they experience embodiment. How they talk also tells us much about the portrayal of bodily experience; that is, the connection between how people talk about their bodies and how they experience them. (Becker, 1997, pp. 29–30)

The cohesive nature of narratives can also be problematic. Such a search for meanings in disruptive events through narrative flow can lead to their fabrication, which may have disastrous results, as is the case for persecutory delusions (Freeman et al., 2002). If the narrative, thus, has the advantage of re-evaluating the individual's experience of the patient for the formulation of a diagnosis, in the same way, the metaphors created in the rhetoric bargaining between doctor and patient are at risk of being exploited to assert cultural and social domination of one of the parties over the other.

1.6 Conclusion

The purpose of this section was to contextualise how the concept of identity has undergone profound changes to reach a complete epistemological revision of the human condition and how the primary mode of expression of such a new condition is through a narrative flow. The distinction previously made between identity's analytical and operational level is essential. On the one hand, identity tends to mean too much when designed as a practice category, especially in the political sense, as part of a language that pushes towards social action. On the other hand, it means little or nothing if approached from a constructivist stance because there is the risk of losing its meaning through excessive theoretical elaboration. The distinction between gender and sex was also taken as an example to reify the malleability of such a concept. In the last few decades, attention has shifted from the biological sex of the individual, conceived as crucial for one's constitution, to the gender, outlined instead in terms of performativity. Gender studies have contributed to a radical reformulation of the human condition, conceived today as much more ambiguous. The idea of a strong association between physical characteristics and personal inclinations is obsolete, at least from an anthropological perspective.

It was chosen to linger on the concept of identity, sex and gender because the given biological body does not always overlap with what the individual deems to be one's identity (the relevance of such an idea and its correlation with postmodernism will be addressed when debating the putative aetiology of BID). How the medical world faces such dilemmas and how the consistency of decisions is applied and maintained is a matter open to debate.

1.7 Book Structure and Necessary Remarks

This book is based on fieldwork and original survey data involving over one thousand physicians across various specialties. The research is shaped by a blend of medical anthropology and bioethics, which has greatly influenced both the theoretical approach and empirical methods throughout this book.

The structure of the book is deliberately organised to reflect its interdisciplinary approach. The theoretical character of Chaps. 1, 2 and 3 sets the ground for an understanding of the complexities surrounding BID and presents a review of the state of research. The empirical findings in Chap. 4 are matched against this analysis, which in turn is furthered in Chaps. 5, 6 and 7, where the empirical data that was examined is put under a magnifying glass. Chapter 8 takes again a more theoretical orientation, allowing for a subtle, philosophical analysis of whether BID amputations can be either justified or refuted based on contemporary philosophical positions. Finally, Chap. 9 synthesises the theoretical and empirical parts by describing their similarities, showing in what ways they reinforce one another, and includes results and personal evaluation of the whole project.

The fieldwork experience contributed significantly to this research, combining interviews and participant observation, focusing mainly on surgeons, as well as psychiatrists and clinical psychologists. In addition, the survey was extended to a broad spectrum of medical doctors specialising in various fields. While the actual research methodology is not discussed in this book, the text makes it clear when one reads insights that are from original data analysis and provides reasoning for the observations made. Because of the complexity of the book at hand and for reasons of space, I only provide passages from the original interviews which have been elaborated through more advanced data analysis. In a few instances, I have kept verbatim quotations because I felt it would be particularly interesting to the reader. When quoting, I have made sure to protect respondent anonymity.

Chapter 1 introduces Body Integrity Dysphoria (BID) as a condition where there is a mismatch between Self and body. Evolving concepts of identity, gender and narrative, particularly about the rise of postmodernism, are receiving new attention against long-established medical and ethical frameworks. The chapter points out the fluidity of identity, emphasising the increasingly blurred line between sex and gender, which creates an interesting starting point to reflect on BID.

Chapter 2 explores the theoretical foundations of BID, focusing on body image and body schema. The chapter argues that such concepts are the result of external stimuli as well as cultural influence, and not only biological ones. It considers the argument that the impairment desire could be potentially motivated by an increasingly fragmented Self due to an image-centric culture. A holistic perspective to understand BID would need to consider arguments within the neurological, psychological and social realms.

Chapter 3 examines the ethical tensions surrounding surgeries for BID. The chapter discusses the ethical issues related to surgeries for BID and compares them with gender-affirmation surgeries. It considers also the bioethical principles of beneficence, non-maleficence and justice, questioning the fairness and accessibility of care. It suggests, moreover, that BID-related amputations should be considered equal to SRS. The chapter does not advocate, however, for such procedures but rather points out the existing difficulties and knowledge gaps within medical practice to identify morally relevant differences between these two practices.

Chapter 4 analyses ethical concerns in cosmetic surgery, focusing on societal beauty norms and the criteria surgeons use to determine the legitimacy of bodily modifications. The chapter examines ideas such as normativity and body harmony, emphasising the role of surgeons in assessing a patient's psychological readiness for surgery. It also addresses the paradox of achieving natural beauty through artificial means, questioning the boundaries of legitimate interventions.

Chapter 5 considers patient competence through pre-operative risk assessment and post-operative behaviour by drawing daring comparisons between gastric banding procedures. It draws parallels with BID, considering how patterns of behaviour and decision-making capacity have bearings on surgical eligibility. It raises critical questions regarding the anticipation of patient regret and the ethics of delaying surgery to consider non-surgical alternatives.

Chapter 6 explores moral responsibility for BID and the need for amputation; whether the person's choice is a product of autonomy or of coercion. The chapter examines the instinctive refusal of on-demand amputations with implications for moral reasoning. Additionally, the ideas of moral agency, distinction between moral and non-moral actions and the influence of socio-cultural contexts on moral cognitions are discussed and applied to BID.

Chapter 7 evaluates the moral implications of BID-related amputation requests, considering whether such actions are praiseworthy or blameworthy. The chapter explores how physicians' preconceptions and scepticism inform their approach to BID; it targets coercion, intention and moral responsibility. The chapter concludes by suggesting that self-inflicted amputations can be commended, in the absence of any effective treatment.

Chapter 8 introduces well-being as a central concept for understanding BID and attempts to provide a model that could support amputation as a valid health-seeking behaviour. The chapter discusses Hedonism, Desire-Fulfilment and Objective-list theories with regard to the benefits of amputation in the case of individuals suffering from BID; such an approach opens different angles on well-being and raises various questions concerning the legitimacy of interventions.

Chapter 9 concludes with a reflective analysis of the ethical and practical challenges surrounding on-demand amputations, summarising the complex issues explored throughout the text.

References

Bayne, T., & Levy, N. (2005). Amputees by choice: Body integrity identity disorder and the ethics of amputation. *Journal of Applied Philosophy, 22*(1), 75–86. https://doi.org/10.1111/j.1468-5930.2005.00293.x

Beauchamp, T. L., & Childress, J. F. (2013). *Principles of biomedical ethics* (7th ed.). Oxford University Press.

Becker, G. (1997). *Disrupted lives: How people create meaning in a chaotic world.* University of California Press.

Bem, S. L. (1974). The measurement of psychological androgyny. *Journal of Consulting and Clinical Psychology, 42*(2), 155–162. https://doi.org/10.1037/h0036215

Bem, S. L. (1981). Gender schema theory: A cognitive account of sex typing. *Psychological Review, 88*(4), 354–364. https://doi.org/10.1037/0033-295X.88.4.354

Berlin, I. (2002). Two concepts of liberty. In *Liberty: Incorporating four essays on liberty* (pp. 166–218). Oxford University Press.

Brubaker, R., & Cooper, F. (2000). Beyond "identity". *Theory and Society, 29*(1), 1–47. https://doi.org/10.1023/A:1007068714468

Buchanan, A., & Brock, D. W. (1986). Deciding for others. *The Milbank Quarterly, 64*(2), 17–94. https://doi.org/10.2307/3349960

Crerand, C. E., Franklin, M. E., & Sarwer, D. B. (2008). MOC-PS(SM) CME article: Patient safety: Body dysmorphic disorder and cosmetic surgery. *Plastic and Reconstructive Surgery, 122*(Supplement), 1–15. https://doi.org/10.1097/PRS.0b013e3181888ffd

David, B., Grace, D., & Ryan, M. K. (2013). The gender wars: A self-categorization theory perspective on the development of gender identity. In M. Bennett & F. Sani (Eds.), *The development of the social self* (Vol. 1, pp. 135–157). Psychology Press.

Freeman, D., Garety, P. A., Kuipers, E., Fowler, D., & Bebbington, P. E. (2002). A cognitive model of persecutory delusions. *British Journal of Clinical Psychology, 41*(4), 331–347. https://doi.org/10.1348/014466502760387461

Harvey, D. (1989). *The condition of postmodernity: An enquiry into the origins of cultural change.* Blackwell.

Kagan, J. (1974). A cognitive-developmental analysis of children's sex-role concepts and attitudes. In M. L. Hoffman (Ed.), *Review of child development research* (Vol. 1, pp. 137–167). Russell Sage Foundation.

Leve, L. (2011). Identity. *Current Anthropology, 52*(4), 513–535. https://doi.org/10.1086/660999

Lurye, L. E., Zosuls, K. M., & Ruble, D. N. (2008). Gender identity and adjustment: Understanding the impact of individual and normative differences in sex typing. *New Directions for Child and Adolescent Development, 2008*(120), 31–46. https://doi.org/10.1002/cd.214

Mead, G. H. (2002). *The philosophy of the present.* Prometheus Books.

Mead, G. H. (2009). *Mind, self, and society: From the standpoint of a social behaviorist* (C. W. Morris, Ed., p. 31). University of Chicago Press.

Nejadsarvari, N., Ebrahimi, A., Ebrahimi, A., & Hashem-Zade, H. (2016). Medical ethics in plastic surgery: A mini review. *World Journal of Plastic Surgery, 5*(3), 207–212.

Powlishta, K. K. (2013). Gender as a social category: Intergroup processes and gender-role development. In M. Bennett & F. Sani (Eds.), *The development of the social self* (Vol. 1, pp. 103–133). Psychology Press.

Robert, G. (1999). The rehabilitation of rehabilitation: A narrative approach to psychosis. In G. Roberts & J. Holmes (Eds.), *Healing stories: Narrative in psychiatry and psychotherapy* (pp. 152–181). Oxford University Press.

Ruble, D. N., Alvarez, J., Bachman, M., Cameron, J., Fuligni, A., Coll, C. G., & Rhee, E. (2013). The development of a sense of 'we': The emergence and implications of children's collective identity. In M. Bennett & F. Sani (Eds.), *The development of the social self* (Vol. 1, pp. 29–76). Psychology Press.

Veatch, R. M., & Guidry-Grimes, L. K. (2020). *The basics of bioethics* (4th ed.). Routledge, Taylor & Francis Group.

Wilson, J. (2007). Is respect for autonomy defensible? *Journal of Medical Ethics, 33*(6), 353–356. https://doi.org/10.1136/jme.2006.018572

Chapter 2
Theoretical Framework: Understanding BID Through Embodiment

2.1 Introduction

Individuals with Body Integrity Dysphoria (BID) experience a disconnect between their physical body and how they perceive it. (First, 2005; First & Fisher, 2012). The BID phenomenon falls under Chap. 6 of the International Classification of Diseases 11th Revision (ICD-11), Code 6C21, which defines BID as "an intense and persistent desire to become physically disabled in a significant way… with onset in early adolescence accompanied by persistent discomfort, or intense feelings of inappropriateness concerning current non-disabled body configuration" (World Health Organization, 2019, 6C21). The inclusion of BID in the ICD-11 is a significant development, as it officially defines and legitimises the condition for the first time. However, this incorporation comes with challenges. While the inclusion of BID in the ICD-11 is a crucial advancement, the understanding of the condition is far from exhaustive. This leaves medical professionals with an officially recognised condition but lacking the means and scientific knowledge to comprehend it and address it fully.

2.2 Three Possible Causes

Neurological analyses of BID individuals suggest an anomaly in the frontoparietal lobe (see among others Brang et al., 2008; Hilti et al., 2013; Van Dijk et al., 2013). Such an alteration would presumably cause dysfunction in both brain structure and function (Blom et al., 2016; Brugger et al., 2016; Hänggi et al., 2016, 2017). This means that individuals with BID show different levels of brain activity connected to the affected limbic regions. Such brain activity appears because perceiving their desired or undesired body state triggers distinct emotional responses or affective

L. Loriga, *Body Integrity Dysphoria and the Ethical Dilemma of On-Demand Amputation*, https://doi.org/10.1007/978-3-031-82443-2_2

states. Interpersonal relationships further influence these neural mechanisms, as sensorimotor processes are impacted by observing and imitating others (Désy & Théoret, 2007; Maister et al., 2013; Tsakiris, 2017b). This conclusion has led to the understanding that the BID spectrum has both neuropsychological and social co-determinants, thereby ruling out pharmacological and neurological surgical approaches as potential treatments (see, for example, the relevance for the BID narrative of the encounters with other amputees or their attraction for the amputee body Aoyama et al., 2012; Brugger et al., 2013; De Preester, 2013; First, 2005). Interestingly, similar factors are reported also as contributing to the increase in cosmetic surgery and related body modification practices (Blond, 2008; Groesz et al., 2002; Mills et al., 2002; Tiggemann et al., 2009).

Conversely, psychological models emphasise a spontaneous and profound fascination with a particular body type (Kasten, 2009; Obernolte et al., 2015). Mental images play a pivotal role here as they entail sensory qualities that can be triggered by external stimuli (Holmes & Mathews, 2010; Kosslyn et al., 2001). This aligns with the narrative of many individuals with BID, who often report that encountering an amputee serves as the focal point in their development of a desire to amputate.

Lastly, a postmodern approach suggests an increasing problematisation of the Self under postmodern conditions due to a growing scepticism towards the fundamental principles of Western thought and the notion of an ultimate, objective truth (Sim, 2011). Whereas pre-modern society offered shared and stable identities through communal bonds (Gross, 1992), modern society shifted away from tradition towards an individualistic conception of the Self (Giddens, 1991), potentially leading to a fragmentation of the individual, who finds themselves in a continuous state of dissolution (Baudrillard, 1994, p. 121). The relevance of the postmodern approach is closely tied to the importance of mental imagery in the BID phenomenon. The growing dependence on an image-centric culture, which is typical of postmodernity, might very well challenge a stable and collective sense of Self. This instability might be seen as a contributing factor to the development of a desire for impairment, similarly to the way the proliferation of media body images has led to an increase in cosmetic and plastic surgery procedures.

2.3 Sex and Gender as a Gateway to Understand BID

The models presented above are not mutually exclusive, as the evidence of structural brain differences related to the desire for impairment does not necessarily point to a purely biological cause (Schwartz et al., 2016). The feeling of dysphoria is also reported as being a causal effect of Gender Dysphoria and cosmetic surgery (Lawrence, 2006, 2009; Mills et al., 2002). Furthermore, the concept of dysphoria seems to imply a differentiation in types of long-term bodily representation (Anema et al., 2009; Botvinick & Cohen, 1998; Carruthers, 2008; Kammers et al., 2009; O'Shaughnessy, 1980, pp. 273–281; Paillard, 2003; Smith, 2009; Vignemont, 2018, pp. 141–163) as well as the idea that such representation can be altered over time

(Cardinali et al., 2009; Lackner, 1988) through sensory and cognitive stimulation (Taylor-Clarke et al., 2004).

To better understand the significance of mismatch in the BID debate, it is useful to examine the evolving differentiation between sex and gender in the International Statistical Classification of Diseases (ICD) by the World Health Organization and the Diagnostic and Statistical Manual of Mental Disorders (DSM) by the American Psychiatric Association. Both have undergone significant revisions. In DSM-5, the previous category of Sexual and Gender Identity Disorders was replaced by Gender Dysphoria (American Psychiatric Association, 2013, p. 451), shifting the focus from cross-gender identification to gender incongruence. Similarly, ICD-11 replaced the terms Gender Identity Disorder and Transsexualism with Gender Incongruence, moving it out of the category of paraphilic disorders. Both DSM-5 and ICD-11 now describe Gender Incongruence as a persistent feeling of mismatch between one's gender identity and biological sex. This reclassification moves the issue from the realm of mental health to sexual health, thus reducing stigma and recognising that one's physical body may not align with one's subjective identity.

2.4 Bodily Representation

The literature on bodily representation models is prolific. Nonetheless, the notion of two distinct representations of one's bodily experiences is prominent and well agreed upon (de Vignemont, 2010; Di Vita et al., 2016; Dijkerman & De Haan, 2007; Dyson et al., 2010; Gallagher, 1986; Hall, 2003; Longo, 2015; Longo et al., 2010; Longo & Haggard, 2012). When taking this latter approach as valid, the term body image refers to a conscious representation of body structure derived from various sensorimotor experiences (Gallagher, 1998, 2001), including localised physical stimuli. In contrast, body schema provides a dynamic, subconscious regulation of posture and movement, enabling the successful performance of actions (Dijkerman & De Haan, 2007; Haggard & Wolpert, 2006). Under typical conditions, body schema and body image function in harmony (Welch & Warren, 1980); in others, they diverge, as might be the case for traumas (Apps & Tsakiris, 2014; Tsakiris, 2017a). When discrepancies occur, the body image recalibrates the body schema, creating a sensation of mismatch. This recalibration can lead to unconventional bodily and perceptual experiences (Begum Ali et al., 2015; Brownell et al., 2007; DeLoache et al., 2004; Martin, 1992).

Individuals within the BID spectrum report a disconnect between their physical bodies and their true selves, similar to the experiences of many transgender individuals (Lindgren & Pauly, 1975; Ramachandran & McGeoch, 2008). In both groups, the desire to transition or amputate often emerges in early childhood (Bruno, 1997; Zucker & Cohen-Kettenis, 2008). Neurological analyses reveal that both Gender Dysphoria and BID individuals exhibit similar physiological arousal when the affected body part is stimulated (Brang et al., 2008; McGeoch et al., 2011). Likewise, both conditions reveal reduced cortical thickness in the right superior parietal lobule (Hilti et al., 2013) indicating the brain's difficulty in integrating sensations from the limb into body maps (Baier & Karnath, 2008; McGeoch et al., 2011; Tsakiris et al., 2007).

A further tempting comparison can also be made with cosmetic and plastic surgery. Plastic surgery typically involves reconstructing or restoring the function of damaged body parts, while cosmetic surgery aims to enhance physical appearance. However, individuals seeking these procedures, unless correcting a physical impairment, might be categorised under Body Dysmorphic Disorder, often invalidating their ability to autonomously decide on undergoing such surgeries (Ackerman, 2010; Morgan, 1991).

Given these premises, one could argue that BID constitutes a similar form of non-delusional somatic alienation as seen sometimes in Gender Dysphoria (Bower, 2001). In contrast to Gender Dysphoria and plastic and cosmetic surgeries, however, BID individuals do not perceive any ugliness or dysfunction in the limb they wish to amputate. At the same time, considerable levels of satisfaction are reported post-amputation (see for example Bayne & Levy, 2005, pp. 80–81). The same satisfaction cannot always be said for Gender Dysphoria patients (Ainsworth & Spiegel, 2010; De Cuypere et al., 2005; de Vries et al., 2014; Gijs & Brewaeys, 2007; Horbach et al., 2015; Kelly et al., 2008; Kuhn et al., 2009; Landén et al., 1998; Lawrence, 2003, 2006; Motmans et al., 2012; Newfield et al., 2006; Peyrot & Rubin, 2009; Pfafflin, 1993; Smith et al., 2005).

2.4.1 *The Idea of Malleability*

For individuals with BID, the sense of a complete Self is not achieved through positive identification with their natural bodily metrics but rather through the absence of a specific body part, suggesting an altered body image. The predominant notion in the BID narrative is a sense of over-completeness, which suggests that the body has physiological characteristics exceeding the individual's internal bodily configuration. Over the years, various definitions of BID have been proposed (Sedda & Bottini, 2014), mainly depending on the investigative approaches employed.

Currently, the emphasis is on the concept of Dysphoria, highlighting a misalignment between one's identity—no longer viewed as dysfunctional—and one's physical body, rather than an identity disorder. This perspective creates a parallel between individuals with BID and those with Gender Dysphoria, who seek Sex Reassignment Surgery (from now on SRS) and are acknowledged as having the medical decision-making capacity to undergo significant surgical interventions to match their physical appearance with their true identity.

The sensation of Dysphoria drives individuals to desire the removal of a specific body part to achieve harmony between their body image and sense of Self. Body image, which involves a conscious awareness of one's body (Di Vita et al., 2016, p. 541) formed from various sensorimotor experiences internalised throughout an individual's life (Gallagher, 1986, pp. 225–244, 1998, pp. 147–176) is central to contemporary understanding of this phenomenon. Alterations in body image are considered to be influenced not only by neurological factors but also by psychological and social elements, such as cultural influences (Freedberg & Pennisi, 2020).

A characteristic feature of BID individuals, in fact, is their encounter with people who have physical impairments, which often catalyses the awareness of their desire for impairment (Obernolte et al., 2015). Such encounters with amputees can also function as retrospective mechanisms, helping individuals justify and trace the development of their impairment desire. This process emphasises the impact of intense emotional experiences on bodily perception (Aoyama et al., 2012). This suggests that there is room for adaptability of long-term bodily representations (Langer et al., 2012).

Traditionally, scientific literature focuses on two different concepts to distinguish between the stable perception that the individual has of one's own body, namely the body image and the variation that such perception assumes to allow the individual to act in the world, body schema. This latter can be regarded as the continuous unconscious adaptation and recalibration of bodily metrics (Gallagher, 1986; O'Shaughnessy, 1980, pp. 282–298; Vignemont, 2018, pp. 140–163). For those experiencing BID, it is crucial to understand whether their body image has been altered and how this potential alteration can be explained through everyday embodied experiences.

Cardinali and colleagues (Cardinali et al., 2009) provide an insightful discussion on how individuals modify their body configuration to complete specific tasks with external devices. They suggest that using tools, such as hammers or brushes—or, in the context of BID, crutches or wheelchairs—alters the body schema not just functionally, but perceptually. This perceptual change indicates that long-term body representations, including body schema and body image, are not fixed but evolve continually, shaped by ongoing life experiences. Studies, like the rubber hand illusion experiments, have shown that bodily sensations can be induced by sensory perceptions beyond mere touch (Botvinick & Cohen, 1998; Ehrsson et al., 2005, 2007).

This proposition is particularly intriguing when applied to BID. It could explain the comparative judgement between oneself and others (Jaegher, 2015), or more specifically, between the BID individual's actual body and their desired impaired body. Through this perspective, body image transcends being solely a neurological phenomenon and becomes a social one. For individuals with BID, this may imply two things: first, the idealisation of an unconventional human body phenotype may stem from a predisposition or overemphatic reaction to the desired impaired body. Second, this predisposition is reinforced through both visual stimuli and pretending behaviour, providing a pleasurable experience by imaginatively omitting the undesired body part (Brugger et al., 2013). Sight therefore plays a crucial role for BID individuals, suggesting the internalisation of a specific mental image associated with the desired state—namely, an impaired body. Through simulation practices, such as pretending behaviours, this mental image is reinforced via positive feedback (Aner et al., 2018; Grocholewski et al., 2018). This alteration in body image subsequently affects the body schema, impacting the individual's situational awareness of the unwanted body part (Hänggi et al., 2016).

2.4.2 The Importance of Society for the Self

Charles Horton Cooley (1983) sees the Self as grounded in three processes: imagining how we appear to others, imagining others' judgements of us and experiencing self-feelings. Similarly, George Herbert Mead situates the Self within the concept of the generalised other (Mead, 2009, pp. 152–163), which involves the internalisation of the individual's entire sociocultural context. This suggests that the Self, along with its bodily representations, is influenced by time and space.

Cultural traditions have historically imposed values, established continuity and codified specific behavioural patterns that supported social existence and understanding of the world. Without a connection to the past, where these values are ideally rooted, the world risks losing its meaning. Consequently, relaxing the societal pressures that are crucial in shaping one's Self can lead to a sense of indeterminacy for the individual. In contemporary society, there has been a notable shift away from traditional values in favour of an ideology of self-reliance (Gross, 1992, pp. 40–91). This transition replaces established meanings with new, individualistic ones that are grounded in personal experiences rather than in collectively held cultural traditions (Frank, 2013, p. 7). The Self becomes "a reflexive project for which the individual is responsible" (Giddens, 1991, p. 75). This shift away from tradition can be attributed mainly to the rise of a visual culture, where images have become the primary mode of communication and an essential commodity. A clear example of this dynamic is the causal relationship between media body images and the increase in cosmetic surgeries (Blond, 2008; Mills et al., 2002). The extensive reliance on visual media as the primary means of communication in the postmodern era has led to images being constructed upon one another without any grounding in tradition or claims to reality. Regarding the body, there has been a shift from a shared notion of what a human body should look like—what Nora L. Jones (Jones, 2011) refers to as the Body as Spectacle, representing the socio-cultural context's standard of bodily characteristics—towards a diversity of bodily representations, each valued equally (Gergen, 1991).

To understand the impact of the postmodern era on individual body image, it is essential to further explore what postmodernity signifies for the body. In the postmodern era, culture is predominantly image-based, with imagery serving as the primary mode of communication. This means also an erosion of traditional or historical meanings, with reality increasingly replaced by the concept of hyperreality—a state where the artificial or simulated becomes more significant than the real. Jean Baudrillard describes this phenomenon as the triumph of the simulacrum, an imitation or representation of something that lacks a true origin or reality, that is "models of a real without origin or reality" (Baudrillard, 1988, p. 166). Consequently, the images that dictate what the individual and the concept of Self should embody begin to replace the actual individual and the authentic Self. In this context, the image itself transforms into a commodity, valued more for its appearance and presentation than for its intrinsic value.

For the BID such speculative reasoning is valuable because of the emphatic response to a particular body type which leads to the internalisation of a specific mental image. This internalisation affects the individual's sensorimotor experience and may result in identifying with a body image that does not align with one's physical metrics (Brugger et al., 2013, p. 5). The significance of mental imagery in BID could be amplified in a society focused on the production of images, as seen in postmodernity. Consequently, such internalisation might not require real-life encounters with impaired-body individuals. It could be reinforced by the absence of a stable self-concept to which the individual can conform.

2.5 Conclusion

This chapter suggested that body image and body schema—conscious bodily awareness and unconscious bodily calibration, respectively—are subject to alteration by various factors beyond biological determinants despite conveying information about the body's enduring properties. An emphatic response to encountering an amputee or internalising specific mental imagery, such as the initial sight of an impaired body, can influence sensorimotor processing and alter body image through positive reinforcement.

Moreover, an intriguing and voluntarily far-fetched and provocative idea has been proposed: the lack of a stable and shared sense of Self, compounded by the dominance of an image-based culture typical of the postmodern condition, may contribute to the desire for impairment, similar to how the rise of media body images led to an increase in cosmetic and plastic surgery procedures.

Currently, neither neurological nor psychological explanations provide a comprehensive account of BID's aetiology. Contemporary formulations of BID (Kasten, 2019) advocate for moving beyond unidimensional explanations and considering neurological, psychological and social factors collectively. Using the concept of embodiment as a central point, this approach aims to develop a more holistic understanding of the desire for impairment.

References

Ackerman, S. L. (2010). Plastic paradise: Transforming bodies and selves in Costa Rica's cosmetic surgery tourism industry. *Medical Anthropology, 29*(4), 403–423. https://doi.org/10.1080/01459740.2010.501316

Ainsworth, T. A., & Spiegel, J. H. (2010). Quality of life of individuals with and without facial feminization surgery or gender reassignment surgery. *Quality of Life Research, 19*(7), 1019–1024. https://doi.org/10.1007/s11136-010-9668-7

American Psychiatric Association (Ed.). (2013). *Diagnostic and statistical manual of mental disorders: DSM-5* (5th ed.). American Psychiatric Association.
Anema, H. A., Van Zandvoort, M. J. E., De Haan, E. H. F., Kappelle, L. J., Kort, D., Paul, L. M., Jansen, B. P. W., & Dijkerman, H. C. (2009). A double dissociation between somatosensory processing for perception and action. *Neuropsychologia, 47*(6), 1615–1620. https://doi.org/10.1016/j.neuropsychologia.2008.11.001
Aner, L.-L., Schmitt, F., & Kasten, E. (2018). The beauty and the BIID. *International Journal of Research in Humanities and Social Studies, 5*(5), 33–53.
Aoyama, A., Krummenacher, P., Palla, A., Hilti, L. M., & Brugger, P. (2012). Impaired spatial-temporal integration of touch in xenomelia (body integrity identity disorder). *Spatial Cognition & Computation, 12*(2–3), 96–110. https://doi.org/10.1080/13875868.2011.603773
Apps, M. A. J., & Tsakiris, M. (2014). The free-energy self: A predictive coding account of self-recognition. *Neuroscience & Biobehavioral Reviews, 41*, 85–97. https://doi.org/10.1016/j.neubiorev.2013.01.029
Baier, B., & Karnath, H.-O. (2008). Tight link between our sense of limb ownership and self-awareness of actions. *Stroke, 39*(2), 486–488. https://doi.org/10.1161/STROKEAHA.107.495606
Baudrillard, J. (1988). Simulacra and simulations. In M. Poster (Ed.), *Jean Baudrillard: Selected writings* (1st ed., pp. 166–184). Stanford University Press.
Baudrillard, J. (1994). *Simulacra and simulation* (S. F. Glaser, Trans.). University of Michigan Press.
Bayne, T., & Levy, N. (2005). Amputees by choice: Body integrity identity disorder and the ethics of amputation. *Journal of Applied Philosophy, 22*(1), 75–86. https://doi.org/10.1111/j.1468-5930.2005.00293.x
Begum Ali, J., Spence, C., & Bremner, A. J. (2015). Human infants' ability to perceive touch in external space develops postnatally. *Current Biology, 25*(20), R978–R979. https://doi.org/10.1016/j.cub.2015.08.055
Blom, R. M., Van Wingen, G. A., Van Der Wal, S. J., Luigjes, J., Van Dijk, M. T., Scholte, H. S., & Denys, D. (2016). The desire for amputation or paralyzation: Evidence for structural brain anomalies in body integrity identity disorder (BIID). *PLoS One, 11*(11), e0165789. https://doi.org/10.1371/journal.pone.0165789
Blond, A. (2008). Impacts of exposure to images of ideal bodies on male body dissatisfaction: A review. *Body Image, 5*(3), 244–250. https://doi.org/10.1016/j.bodyim.2008.02.003
Botvinick, M., & Cohen, J. (1998). Rubber hands 'feel' touch that eyes see. *Nature, 391*(6669), 756–756. https://doi.org/10.1038/35784
Bower, H. (2001). The gender identity disorder in the DSM-IV classification: A critical evaluation. *Australian & New Zealand Journal of Psychiatry, 35*(1), 1–8. https://doi.org/10.1046/j.1440-1614.2001.00859.x
Brang, D., McGeoch, P. D., & Ramachandran, V. S. (2008). Apotemnophilia: A neurological disorder. *Neuroreport, 19*(13), 1305–1306. https://doi.org/10.1097/WNR.0b013e32830abc4d
Brownell, C. A., Zerwas, S., & Ramani, G. B. (2007). "So big": The development of body self-awareness in toddlers. *Child Development, 78*(5), 1426–1440. https://doi.org/10.1111/j.1467-8624.2007.01075.x
Brugger, P., Lenggenhager, B., & Giummarra, M. J. (2013). Xenomelia: A social neuroscience view of altered bodily self-consciousness. *Frontiers in Psychology, 4*. https://doi.org/10.3389/fpsyg.2013.00204
Brugger, P., Christen, M., Jellestad, L., & Hänggi, J. (2016). Limb amputation and other disability desires as a medical condition. *The Lancet Psychiatry, 3*(12), 1176–1186. https://doi.org/10.1016/S2215-0366(16)30265-6
Bruno, R. L. (1997). Devotees, pretenders and wannabes: Two cases of factitious disability disorder. *Sexuality and Disability, 15*(4), 243–260. https://doi.org/10.1023/A:1024769330761
Cardinali, L., Frassinetti, F., Brozzoli, C., Urquizar, C., Roy, A. C., & Farnè, A. (2009). Tool-use induces morphological updating of the body schema. *Current Biology, 19*(12), R478–R479. https://doi.org/10.1016/j.cub.2009.05.009

Carruthers, G. (2008). Types of body representation and the sense of embodiment. *Consciousness and Cognition, 17*(4), 1302–1316. https://doi.org/10.1016/j.concog.2008.02.001

Cooley, C. H. (1983). The social self – the meaning of I. In *Human nature and the social order* (pp. 88–109). Transaction Books.

De Cuypere, G., et al. (2005). Sexual and physical health after sex reassignment surgery. *Archives of Sexual Behavior, 34*(6), 679–690. https://doi.org/10.1007/s10508-005-7926-5

De Preester, H. (2013). Merleau-Ponty's sexual schema and the sexual component of body integrity identity disorder. *Medicine, Health Care and Philosophy, 16*(2), 171–184. https://doi.org/10.1007/s11019-011-9367-3

de Vignemont, F. (2010). Body schema and body image – Pros and cons. *Neuropsychologia, 48*(3), 669–680. https://doi.org/10.1016/j.neuropsychologia.2009.09.022

de Vignemont, F. (2018). Taxonomies of body representations. In *Mind the body: An exploration of bodily self-awareness* (1st ed., pp. 140–163). Oxford University Press.

de Vries, A. L. C., McGuire, J. K., Steensma, T. D., Wagenaar, E. C. F., Doreleijers, T. A. H., & Cohen-Kettenis, P. T. (2014). Young adult psychological outcome after puberty suppression and gender reassignment. *Pediatrics, 134*(4), 696–704. https://doi.org/10.1542/peds.2013-2958

DeLoache, J. S., Uttal, D. H., & Rosengren, K. S. (2004). Scale errors offer evidence for a perception-action dissociation early in life. *Science, 304*(5673), 1027–1029. https://doi.org/10.1126/science.1093567

Désy, M.-C., & Théoret, H. (2007). Modulation of motor cortex excitability by physical similarity with an observed hand action. *PLoS One, 2*(10), e971. https://doi.org/10.1371/journal.pone.0000971

Di Vita, A., Boccia, M., Palermo, L., & Guariglia, C. (2016). To move or not to move, that is the question! Body schema and non-action oriented body representations: An fMRI meta-analytic study. *Neuroscience & Biobehavioral Reviews, 68*, 37–46. https://doi.org/10.1016/j.neubiorev.2016.05.005

Dijkerman, H. C., & De Haan, E. H. F. (2007). Somatosensory processes subserving perception and action. *Behavioral and Brain Sciences, 30*(2), 189–201. https://doi.org/10.1017/S0140525X07001392

Dyson, B. J., Dunn, A. K., & Alain, C. (2010). Ventral and dorsal streams as modality-independent phenomena. *Cognitive Neuroscience, 1*(1), 64–65. https://doi.org/10.1080/17588920903501966

Ehrsson, H. H., Holmes, N. P., & Passingham, R. E. (2005). Touching a rubber hand: Feeling of body ownership is associated with activity in multisensory brain areas. *The Journal of Neuroscience, 25*(45), 10564–10573. https://doi.org/10.1523/JNEUROSCI.0800-05.2005

Ehrsson, H. H., Wiech, K., Weiskopf, N., Dolan, R. J., & Passingham, R. E. (2007). Threatening a rubber hand that you feel is yours elicits a cortical anxiety response. *Proceedings of the National Academy of Sciences, 104*(23), 9828–9833. https://doi.org/10.1073/pnas.0610011104

First, M. B. (2005). Desire for amputation of a limb: Paraphilia, psychosis, or a new type of identity disorder. *Psychological Medicine, 35*(6), 919–928. https://doi.org/10.1017/S0033291704003320

First, M. B., & Fisher, C. E. (2012). Body integrity identity disorder: The persistent desire to acquire a physical disability. *Psychopathology, 45*(1), 3–14. https://doi.org/10.1159/000330503

Frank, A. W. (2013). *The wounded storyteller: Body, illness, and ethics* (2nd ed.). The University of Chicago Press.

Freedberg, D., & Pennisi, A. (2020). The body in the picture. The lesson of phantom limbs and the origins of the BIID. *Reti, Saperi, Linguaggi, 9*(17), 5–50. https://doi.org/10.12832/98417

Gallagher, S. (1986). Body image and body schema: A conceptual clarification. *The Journal of Mind and Behavior, 7*(4), 541–554.

Gallagher, S. (1998). Body schema and intentionality. In J. L. Bermúdez, N. Eilan, & A. Marcel (Eds.), *The body and the self* (pp. 225–244). The MIT Press. https://doi.org/10.7551/mitpress/1640.001.0001

Gallagher, S. (2001). Dimensions of embodiment: Body image and body schema in medical contexts. In S. K. Toombs (Ed.), *Handbook of phenomenology and medicine* (Vol. 68, pp. 147–176). Springer. https://doi.org/10.1007/978-94-010-0536-4

Gergen, K. J. (1991). The self under siege. In *The saturated self: Dilemmas of identity in contemporary life* (pp. 1–17). Basic Books.

Giddens, A. (1991). The trajectory of the self. In *Modernity and self-identity: Self and society in the late modern age* (pp. 70–109). Stanford University Press.

Gijs, L., & Brewaeys, A. (2007). Surgical treatment of gender dysphoria in adults and adolescents: Recent developments, effectiveness, and challenges. *Annual Review of Sex Research, 18*(1), 178–224. https://doi.org/10.1080/10532528.2007.10559851

Grocholewski, A., Bücker, C., Kasten, E., Osen, B., & Heinrichs, N. (2018). Mental images in body integrity identity disorder (BIID) and in Body dysmorphic disorder (BDD): An exploratory study. *Behaviour Change, 35*(3), 174–184. https://doi.org/10.1017/bec.2018.17

Groesz, L. M., Levine, M. P., & Murnen, S. K. (2002). The effect of experimental presentation of thin media images on body satisfaction: A meta-analytic review. *International Journal of Eating Disorders, 31*(1), 1–16. https://doi.org/10.1002/eat.10005

Gross, D. (1992). Tradition under stress. In *The past in ruins: Tradition and the critique of modernity* (pp. 20–40). University of Massachusetts Press.

Haggard, P., & Wolpert, D. M. (2006). Disorders of body schema. In H.-J. Freund, M. Jeannerod, M. Hallett, & R. Leiguarda (Eds.), *Higher-order motor disorders: From neuroanatomy and neurobiology to clinical neurology* (pp. 261–272). Oxford University PressOxford. https://doi.org/10.1093/oso/9780198525769.003.0014

Hall, D. A. (2003). Auditory pathways: Are 'What' and 'Where' appropriate? *Current Biology, 13*(10), R406–R408. https://doi.org/10.1016/S0960-9822(03)00323-3

Hänggi, J., Bellwald, D., & Brugger, P. (2016). Shape alterations of basal ganglia and thalamus in xenomelia. *NeuroImage: Clinical, 11*, 760–769. https://doi.org/10.1016/j.nicl.2016.05.015

Hänggi, J., Vitacco, D. A., Hilti, L. M., Luechinger, R., Kraemer, B., & Brugger, P. (2017). Structural and functional hyperconnectivity within the sensorimotor system in xenomelia. *Brain and Behavior, 7*(3), e00657. https://doi.org/10.1002/brb3.657

Hilti, L. M., Hänggi, J., Vitacco, D. A., Kraemer, B., Palla, A., Luechinger, R., Jäncke, L., & Brugger, P. (2013). The desire for healthy limb amputation: Structural brain correlates and clinical features of xenomelia. *Brain, 136*(1), 318–329. https://doi.org/10.1093/brain/aws316

Holmes, E. A., & Mathews, A. (2010). Mental imagery in emotion and emotional disorders. *Clinical Psychology Review, 30*(3), 349–362. https://doi.org/10.1016/j.cpr.2010.01.001

Horbach, S. E. R., Bouman, M.-B., Smit, J. M., Özer, M., Buncamper, M. E., & Mullender, M. G. (2015). Outcome of vaginoplasty in male-to-female transgenders: A systematic review of surgical techniques. *The Journal of Sexual Medicine, 12*(6), 1499–1512. https://doi.org/10.1111/jsm.12868

Jaegher, H. D. (2015). How we affect each other. Michel Henry's 'pathos-with' and the enactive approach to intersubjectivity. *Journal of Consciousness Studies, 22*(1–2), 112–132.

Jones, N. L. (2011). Embodied ethics: From the body as specimen and spectacle to the body as patient. In *A companion to the anthropology of the body and embodiment* (pp. 72–85). Wiley-Blackwell.

Kammers, M. P. M., De Vignemont, F., Verhagen, L., & Dijkerman, H. C. (2009). The rubber hand illusion in action. *Neuropsychologia, 47*(1), 204–211. https://doi.org/10.1016/j.neuropsychologia.2008.07.028

Kasten, E. (2009). Body integrity identity disorder (BIID): Befragung von Betroffenen und Erklärungsansätze. *Fortschritte der Neurologie Psychiatrie, 77*(01), 16–24. https://doi.org/10.1055/s-0028-1100837

Kasten, E. (2019). Which leg shall I cut off? Body Integrity Dysphoria (BID) is not only a neurologic disease: Results of five case-studies. *Case Studies Journal, 8*(12), 31–34.

Kelly, M. M., Tyrka, A. R., Price, L. H., & Carpenter, L. L. (2008). Sex differences in the use of coping strategies: Predictors of anxiety and depressive symptoms. *Depression and Anxiety, 25*(10), 839–846. https://doi.org/10.1002/da.20341

Kosslyn, S. M., Ganis, G., & Thompson, W. L. (2001). Neural foundations of imagery. *Nature Reviews Neuroscience, 2*(9), 635–642. https://doi.org/10.1038/35090055

Kuhn, A., Bodmer, C., Stadlmayr, W., Kuhn, P., Mueller, M. D., & Birkhäuser, M. (2009). Quality of life 15 years after sex reassignment surgery for transsexualism. *Fertility and Sterility, 92*(5), 1685–1689. https://doi.org/10.1016/j.fertnstert.2008.08.126

Lackner, J. R. (1988). Some proprioceptive influences on the perceptual representation of body shape and orientation. *Brain, 111*(2), 281–297. https://doi.org/10.1093/brain/111.2.281

Landén, M., Wålinder, J., Hambert, G., & Lundström, B. (1998). Factors predictive of regret in sex reassignment. *Acta Psychiatrica Scandinavica, 97*(4), 284–289. https://doi.org/10.1111/j.1600-0447.1998.tb10001.x

Langer, N., Hänggi, J., Müller, N. A., Simmen, H. P., & Jäncke, L. (2012). Effects of limb immobilization on brain plasticity. *Neurology, 78*(3), 182–188. https://doi.org/10.1212/WNL.0b013e31823fcd9c

Lawrence, A. A. (2003). Factors associated with satisfaction or regret following male-to-female sex reassignment surgery. *Archives of Sexual Behavior, 32*(4), 299–315. https://doi.org/10.1023/A:1024086814364

Lawrence, A. A. (2006). Patient-reported complications and functional outcomes of male-to-female sex reassignment surgery. *Archives of Sexual Behavior, 35*(6), 717–727. https://doi.org/10.1007/s10508-006-9104-9

Lawrence, A. A. (2009). Parallels between gender identity disorder and body integrity identity disorder: A review and update. In A. Stirn, A. Thiel, & S. Oddo (Eds.), *Body integrity identity disorder: Psychological, neurobiological, ethical and legal aspects* (pp. 154–174). Pabst.

Lindgren, T. W., & Pauly, I. B. (1975). A body image scale for evaluating transsexuals. *Archives of Sexual Behavior, 4*(6), 639–656. https://doi.org/10.1007/BF01544272

Longo, M. R. (2015). Implicit and explicit body representations. *European Psychologist, 20*(1), 6–15. https://doi.org/10.1027/1016-9040/a000198

Longo, M. R., & Haggard, P. (2012). Implicit body representations and the conscious body image. *Acta Psychologica, 141*(2), 164–168. https://doi.org/10.1016/j.actpsy.2012.07.015

Longo, M. R., Azañón, E., & Haggard, P. (2010). More than skin deep: Body representation beyond primary somatosensory cortex. *Neuropsychologia, 48*(3), 655–668. https://doi.org/10.1016/j.neuropsychologia.2009.08.022

Maister, L., Sebanz, N., Knoblich, G., & Tsakiris, M. (2013). Experiencing ownership over a dark-skinned body reduces implicit racial bias. *Cognition, 128*(2), 170–178. https://doi.org/10.1016/j.cognition.2013.04.002

Martin, M. (1992). Sight and touch. In T. Crane (Ed.), *The contents of experience: Essays on perception* (pp. 196–215). Cambridge University Press.

McGeoch, P. D., Brang, D., Song, T., Lee, R. R., Huang, M., & Ramachandran, V. S. (2011). Xenomelia: A new right parietal lobe syndrome. *Journal of Neurology, Neurosurgery & Psychiatry, 82*(12), 1314–1319. https://doi.org/10.1136/jnnp-2011-300224

Mead, G. H. (2009). *Mind, self, and society: From the standpoint of a social behaviorist* (C. W. Morris, Ed., p. 31). University of Chicago Press.

Mills, J. S., Polivy, J., Herman, C. P., & Tiggemann, M. (2002). Effects of exposure to thin media images: Evidence of self-enhancement among restrained eaters. *Personality and Social Psychology Bulletin, 28*(12), 1687–1699. https://doi.org/10.1177/014616702237650

Morgan, K. P. (1991). Women and the knife: Cosmetic surgery and the colonization of women's bodies. *Hypatia, 6*(3), 25–53. https://doi.org/10.1111/j.1527-2001.1991.tb00254.x

Motmans, J., Meier, P., Ponnet, K., & T'Sjoen, G. (2012). Female and male transgender quality of life: Socioeconomic and medical differences. *The Journal of Sexual Medicine, 9*(3), 743–750. https://doi.org/10.1111/j.1743-6109.2011.02569.x

Newfield, E., Hart, S., Dibble, S., & Kohler, L. (2006). Female-to-male transgender quality of life. *Quality of Life Research, 15*(9), 1447–1457. https://doi.org/10.1007/s11136-006-0002-3
O'Shaughnessy, B. (1980). The body-image (2): The long-term body-image. In *The will: Volume 1, dual aspect theory* (pp. 282–269). Cambridge University Press.
Obernolte, C., Schnell, T., & Kasten, E. (2015). The role of specific experiences in childhood and youth in the development of body integrity identity disorder (BIID). *American Journal of Applied Psychology, 4*(1), 1–8. https://doi.org/10.11648/j.ajap.20150401.11
Paillard, J. (2003). Body schema and body image – A double dissociation in deafferented patients. In N. Gantchev, S. Mori, & J. Massion (Eds.), *Motor control, today and tomorrow* (pp. 197–215).
Peyrot, M., & Rubin, R. R. (2009). How does treatment satisfaction work?: Modeling determinants of treatment satisfaction and preference. *Diabetes Care, 32*(8), 1411–1417. https://doi.org/10.2337/dc08-2256
Pfafflin, F. (1993). Regrets after sex reassignment surgery. *Journal of Psychology & Human Sexuality, 5*(4), 69–85. https://doi.org/10.1300/J056v05n04_05
Ramachandran, V. S., & McGeoch, P. D. (2008). Phantom penises in transsexuals. *Journal of Consciousness Studies, 15*(1), 5–16.
Schwartz, S. J., Lilienfeld, S. O., Meca, A., & Sauvigné, K. C. (2016). The role of neuroscience within psychology: A call for inclusiveness over exclusiveness. *American Psychologist, 71*(1), 52–70. https://doi.org/10.1037/a0039678
Sedda, A., & Bottini, G. (2014). Apotemnophilia, body integrity identity disorder or xenomelia? Psychiatric and neurologic etiologies face each other. *Neuropsychiatric Disease and Treatment, 10*, 1255–1265. https://doi.org/10.2147/NDT.S53385
Sim, S. (2011). Postmodernism and philosophy. In S. Sim (Ed.), *The Routledge companion to postmodernism* (3rd ed., pp. 3–15). Routledge.
Smith, A. J. T. (2009). Acting on (bodily) experience. *PSYCHE: An Interdisciplinary Journal of Research on Consciousness, 15*(1), 82–99.
Smith, Y. L. S., Van Goozen, S. H. M., Kuiper, A. J., & Cohen-Kettenis, P. T. (2005). Sex reassignment: Outcomes and predictors of treatment for adolescent and adult transsexuals. *Psychological Medicine, 35*(1), 89–99. https://doi.org/10.1017/S0033291704002776
Taylor-Clarke, M., Jacobsen, P., & Haggard, P. (2004). Keeping the world a constant size: Object constancy in human touch. *Nature Neuroscience, 7*(3), 219–220. https://doi.org/10.1038/nn1199
Tiggemann, M., Polivy, J., & Hargreaves, D. (2009). The Processing of Thin Ideals in Fashion Magazines: A Source of Social Comparison or Fantasy? *Journal of Social and Clinical Psychology, 28*(1), 73–93. https://doi.org/10.1521/jscp.2009.28.1.73
Tsakiris, M. (2017a). The material me: Unifying the exteroceptive and interoceptive sides of the bodily self. In F. de Vignemont & A. J. T. Alsmith (Eds.), *The subject's matter: Self-consciousness and the body* (pp. 365–382). MIT Press.
Tsakiris, M. (2017b). The multisensory basis of the self: From body to identity to others. *Quarterly Journal of Experimental Psychology, 70*(4), 597–609. https://doi.org/10.1080/17470218.2016.1181768
Tsakiris, M., Hesse, M. D., Boy, C., Haggard, P., & Fink, G. R. (2007). Neural signatures of body ownership: A sensory network for bodily self-consciousness. *Cerebral Cortex, 17*(10), 2235–2244. https://doi.org/10.1093/cercor/bhl131
Van Dijk, M. T., Van Wingen, G. A., Van Lammeren, A., Blom, R. M., De Kwaasteniet, B. P., Scholte, H. S., & Denys, D. (2013). Neural basis of limb ownership in individuals with body integrity identity disorder. *PLoS One, 8*(8), e72212. https://doi.org/10.1371/journal.pone.0072212
Welch, R. B., & Warren, D. H. (1980). Immediate perceptual response to intersensory discrepancy. *Psychological Bulletin, 88*(3), 638–667. https://doi.org/10.1037/0033-2909.88.3.638

World Health Organization. (2019). *International statistical classification of diseases and related health problems* (11th ed.). Retrieved January 6, 2024, from https://icd.who.int/en

Zucker, K. J., & Cohen-Kettenis, P. T. (2008). Gender identity disorder in children and adolescents. In D. Rowland & L. Incrocci (Eds.), *Handbook of sexual and gender identity disorders* (pp. 376–422). Wiley.

Chapter 3
The Ethics of Body-Modifying Surgeries

3.1 Introduction

Body Integrity Dysphoria (BID) individuals who successfully managed to undergo amputation procedures report high satisfaction (Blom et al., 2012, 2014). Surgery is far less expensive for the health service provider than long-term psychological or pharmacological treatment (see, for example, Druss et al., 2002). However, amputation surgery is not reversible. The BID phenomenon offers a particularly fertile environment for bioethical elaboration. It represents an intriguing dilemma on the contextual aspect of bioethics approaches and their application in cases where potential underlying psychological causes are addressed through irreversible medical body-modifying procedures (Beckford-Ball, 2000). The main concerns with indulging in such procedures stem from the problem of surgical treatments to resolve alleged and potential psychiatric disorders (see, for example, the work of Engelhardt, 1974 on the medicalisation of masturbatory practices through history). Accepting surgery as a treatment for the BID phenomenon might be seen as a gateway to justify self-harm; hence, the decision to indulge in such a procedure needs to be pondered carefully. The first issue encountered when approaching the BID phenomenon is the tension between the concepts of harm, beneficence, non-maleficence and autonomy (Gordon et al., 2011).

This section sets the foundations for an ethical analysis of BID, focussing on the tension between beneficence and non-maleficence. The central question is whether fulfilling the desire for amputation enhances well-being or not. The focus is then moved towards the principle of justice, drawing comparisons between BID and Gender Dysphoria. By highlighting parallels between BID's amputation desire and Sex Reassignment Surgeries (SRS), the following argument is made; there is no morally significant distinction for accepting SRS and rejecting BID. Such an argument is voluntarily provocative and it is used to reflect on topics such as fairness, equality and access to care for individuals with BID.

L. Loriga, *Body Integrity Dysphoria and the Ethical Dilemma of On-Demand Amputation*, https://doi.org/10.1007/978-3-031-82443-2_3

3.2 Why the Principle-Based Approach

Before proceeding, it is necessary to clarify the ethical framework that will guide the assessment. While not the only approach, the four-principles model of medical ethics—autonomy, beneficence, non-maleficence and justice—is highly influential today and serves as a foundational framework for understanding ethical considerations. (Gillon, 2003). It is an ideal candidate. While various scholars have criticised the four-principles model (Harris, 2003; Takala, 2001), it remains a widely used and debated framework in both practical and academic settings. The central premise of the four-principles model is that no single principle automatically outweighs the others. Instead, principlism demands a balanced assessment of the specific circumstances surrounding each ethical dilemma.

Historically, the model initially emphasised beneficence and non-maleficence (Smith, 2005). However, over time, there has been a significant shift towards recognising the importance of autonomy as a central ethical principle (Gillon, 1994). Autonomy, which concerns an individual's capacity to make informed decisions about their own lives, has gained prominence in ethical discussions and is now seen as a crucial component of patient-centred care. This shift towards autonomy reflects a changing understanding of how the patient-doctor relationship is perceived. Although beneficence and non-maleficence remain critical aspects, the increased focus on autonomy represents a move towards a more patient-centred approach that values individuals' perspectives and preferences.

3.3 Beneficence

The concept of beneficence can be understood as the proactive promotion of good. In the medical world, beneficence postulates that it is the physician's responsibility to promote the patients' physical and psychological health by considering their best interests and removing any possible harm.

This principle is a prima facie moral obligation for the medical world, which means, however, that its applicability is not universal; in other words, the principle of beneficence is not absolute and can be overridden when it conflicts with an equal or a more vital obligation. Another point to note is that the principle of beneficence is not exclusive, which means that beneficence is a concept that the physician, or someone on their behalf, needs to apply to the general public. Such acceptance of beneficence implies that the physician can and shall choose to prioritise societal well-being if and when recognised that reaching patients' best interest might be detrimental to the broader socio-cultural context. When such understanding is used to frame the BID phenomenon, it can lead to seeing amputation requests as a way to create a precedent that might be used to escalate new claims of a similar nature that would eventually derail society's moral stance. This is a slippery slope type of argument, which, despite its inherent fallacies, has some validity. On the other hand, prioritising the general public's well-being over a single patient can create the same turmoil (Allard & Fortin, 2017; Jeong et al., 2020; Lesieur et al., 2013; Suntharalingam et al., 2009). Additionally, sanctioning amputation procedures as a treatment might give rise to a looping effect (Hacking, 2006, 2007), where the diagnosis assesses and defines the identities, not vice versa. In other words, the awareness of the disorder and the availability of a recognised medical treatment might contribute to the proliferation of the condition. This is to remark that justifying BID by promoting the patient's psychosomatic well-being is not convincing enough. No narrative, either BID, Gender Dysphoria, Cosmetic or Plastic surgery, will suffice if analysed only from a patient-centred perspective. In other words, biomedical cases should be deliberated case-by-case based on the circumstances, including geographical, social and political context. However, from a physician-patient relationship point of view, things change significantly. Any individual who acts autonomously and is recognised to have medical decision-making capacity (Appelbaum, 2007; Appelbaum & Grisso, 1988; Darby & Dickerson, 2017) ought to have their decisions and wishes given due weight, mainly if there are precedents that demonstrate that they are likely to relieve the subject from a state of suffering. The starting point of such speculation is the suffering caused by the mismatch between the perceived body and the actual body configuration. This suffering creates an obligation of beneficence, as there are actions with a high probability of preventing further harm that does not present significant risk, cost or burden if performed under diligent medical attention. In short, the benefits outweigh the risks. While this may hold for the individual making the request, it might be different for the administering entity, which could face legal challenges despite the autonomy and decisional capacity of the individual requesting the intervention. There are limits to what an autonomous agent can do. Certain things are unalienable. The issue with BID, moreover, might not primarily be the amputation of a healthy body part, as this procedure is accepted in other contexts, such as plastic and cosmetic surgeries, gender-affirming surgeries and living-donor organ transplantation. The issue seems to point towards medical malpractice and criminal allegations, as amputation is not yet recognised as the standard treatment for this new medical condition (Johnston & Elliott, 2002).

3.4 Non-maleficence

Beneficence is intrinsically tied to action. This is to say that a physician shall act in those instances in which such a principle calls for action, having in mind the best interest of the patient, and shall not act in all those instances in which such a principle does not call for action. Although such a principle seems straightforward, it is nevertheless possible to speculate on the blurry line separating good and bad outcomes. Suppose an unconscious individual is bleeding to death and is rushed to the Intensive Care Unit (ICU). Without any further information on such an individual, the physician on duty will likely perform all necessary actions to identify and stop the bleeding before replacing the blood lost through a transfusion to stabilise one's condition. Such a situation does not present any particular moral or ethical concerns.

Suppose now there is a similar scenario with the only difference that the physician knows that the individual rushed to the ICU, if conscious, would surely refuse any blood transfusion on a religious basis (see, for example, literature on similar instances for Jehovah's Witnesses Chand et al., 2014). Suppose that no blood substitute is available and that the only way to save the patient's life is to stabilise their condition through the blood available on site. In the first instance, things seem straightforward; to save one's life, the physician needs to perform a series of actions, all of which fall clearly under the spectrum of the principle of beneficence. In the second instance, however, there is an evident clash with the principle of non-maleficence as the physician is forced to make a call to save the patient's life through transfusion, knowing that this action will go against the patient's best interest and might seriously hinder their life choices, or to let the patient die, thus respecting their wishes. The core of the issue seems to be what is considered more harmful in that particular instance, the patient's death (as the preferred outcome per themselves and their family's wishes for such instances) or the patient's life as a possible outcast from the sociocultural and religious setting of belonging. Although the concept of harm is paramount in assessing such situations, it will not be addressed here (but will be done so in the following sections). Instead, what is possible for the time being is to transpose such a dilemma, with the necessary amendments, to the BID phenomenon.

Suppose now a scenario not involving (necessarily) the patient's death as a direct result of the physician's inactivity. Instead, a scenario in which the patient's well-being can (potentially) experience a sharp spike for the better depending on the physician's activity. This means that the physician has the means necessary to relieve the patient from dissatisfaction and suffering through action. Suppose now that such a situation is lifted and shifted to the BID case. The physician has the means necessary, that is, the medical skillset needed to perform amputation. While the logic applied to the ICU example also seems valid for the BID example, the same fundamental issue remains: What defines non-maleficence for the patient? What defines non-maleficence for the physician? Must these two views be separate concepts, or can they converge, and if so, where?

3.5 Autonomy

The concept of autonomy in moral philosophy and bioethics recognizes the human capacity for self-determination, and puts forward a principle that the autonomy of persons ought to be respected. At this level of generality, there is not much with which to take issue; a full account of autonomy must further define self-determination and state how and to what extent autonomy should be respected. (Reich, 1995, p. 246).

In the literature on elective body-modifying surgeries, the four-principal approach is not given equal attention, with the principle of autonomy being far more present and discussed (Chung et al., 2009). The bioethical concept of autonomy "…can be described as a high level of self-determination that few persons will actually achieve, and yet it can still be regarded as a capacity for all persons, if it is believed that all persons under suitable conditions could acquire it and use it to direct their lives" (Reich, 1995, p. 248). Patient autonomy encompasses both negative and positive rights. Negative rights refer to the individual's freedom from external coercion or interference by others. Positive rights involve the individual's ability to exercise self-determination and make informed choices regarding their healthcare (Cohen, 2019; Pellegrino & Thomasma, 1987). Once the informed consent process is completed, any legally competent individual has the right to decide whether to undergo medical procedures. When a patient is determined to have the capacity to make informed decisions, their preferences and choices must be respected. In medical decision-making, an individual's understanding of beneficence should also be taken into account. The challenge here lies in the epistemological aspect of the autonomy principle: evaluating the individual's capacity and ensuring that their autonomy is respected. It is possible that individuals with BID experience monothematic delusions (Davies et al., 2001; Dudley et al., 1997), which could lead to concerns about decisional incapacity, as their desires may originate from an irrational mental state. A delusion is "fixed beliefs that are not amenable to change in light of conflicting evidence" (American Psychiatric Association, 2013, p. 87), with a monothematic delusion being "present in isolation in people whose beliefs are otherwise entirely unremarkable" (Coltheart et al., 2007, p. 1) as might be the limb to be removed for BID. Whether such an argument seems to stand on solid grounds, it is not as one might argue that although the amputation beliefs arrive from an irrational state, the patient's considerations on what to do with it and how to act are rational and ought to be considered. This approach could create a perpetual cycle where weakened practical reasoning perpetuates irrational claims. Conversely, arguments for rationality are often supported by the idea that very few situations can be deemed entirely rational. An interesting perspective on approaching such a dilemma within medical practice is to question whether the will of bodily modification conforms to the individual's second-order volition (Frankfurt, 1971); in other words, whether the individual's will is a free expression of one's decisional capacity. A behaviour limited to acting without coercion cannot be considered free will, but to be such, it requires a certain degree of control over the desires and the actions that arise from them. A subject, therefore, is endowed with free will when they can desire which desires to express, and actually act on them by expressing them in the outside world.

To fully understand the concept of second-order volition, it is necessary to compare and define the latter with two other types of desire; the first-order desire, that is, desire inherent to a direct action as might be in eating when hungry or quenching thirst if thirsty, and the second-order desire, which precedes but does not constrain the development of the second-order volition, which in turn implies the ability to desire or refraining from desiring a specific type of desire. Given these premises, it is possible to distinguish between two second-order desires; the second-order desire translates into the desire to have a particular type of first-order desire for the realisation of which one can refrain from action. This type of desire does not imply yet free will. Only with second-order volition is the individual's free will expressed. Therefore, the fundamental difference between second-order desire and second-order volition lies in the fact that second-order volition is a second-order desire inherent to a first-order desire which motivates the subject to action (Frankfurt, 1971, p. 10).

Therefore, it is possible to speak of free will when there is an overlap between the first-order desire and the second-order desire or when the desire that one wants to motivate an action ultimately brings the subject to action. Therefore, while a person may be free from external constraints, true freedom requires a second-order desire that aligns with a first-order desire motivating their actions. Things get complicated when there is a conflict between second-order desires inherent to first-order desires. In such cases, the individual's freedom is lost as there are no longer second-order desires to which first-order desires can comply. Such reasoning can be translated into the following: a BID individual can be considered free if, and only if, one's will corresponds with what one wants indeed to do. A BID individual, therefore, is free from internal constriction if, and only if, one's desire to amputate corresponds with one's will to want to amputate. Conversely, if the desire for impairment goes against the individual's will to wish for something else, then the freedom of the will is impacted. Whether it is, therefore, legitimate to speak of the principle of alternate possibilities (Frankfurt, 1969), the question inherent to the degree of freedom of the BID remains in the same fashion as it does for an individual who falls under the Gender Dysphoria spectrum as well as for those subjects asking for Cosmetic or Plastic surgery interventions. The pivotal point is whether the individual could have done something other than what they did. Indulging in such speculation sets the bar of free will and autonomous consent apparently too high as it would require a regression of motives virtually infinite. Similarly, it is reasonable to consider how the desire for bodily modification might be a result of psychological and emotional suffering. In such circumstances, one might question how this emotional state impacts the individual's decisional capacity. When applying the emotional narrative to body modification desires, one might wonder how the individual can make rational decisions about achieving a new physical and psychological state that cannot be experienced a priori. In other words, these individuals may not be able to provide informed consent because they cannot fully understand what it would be like to have the body they idealise. Furthermore, their desire for bodily modification might stem from a need to escape their current state of suffering. This does not necessarily mean that the actual modifications they seek will improve their psychological well-being. Once again, indulging in such speculation sets the bar too high for free will and autonomous consent; informed consent entails an adequate understanding of an action's likely consequences.

3.6 Justice

Justice is "the appropriate treatment in light of what is due or owed to a person" (Beauchamp & Childress, 2013, p. 226). Differences can be justified only if they are based on ethically and morally significant distinctions among individuals (Beauchamp & Childress, 2013, p. 262). For the medical setting this means that each individual has the right to seek and request medical treatment to alleviate their suffering. The challenge lies in determining what should be provided and under what specific conditions.

At this point, few questions arise: If dysphoria is now the predominant characterisation of BID, as recognised in the ICD-11, why is the condition not treated similarly to other conditions where dysphoria is the primary feature? Additionally, why is there a discrepancy in the medical approach and availability of treatment options for BID compared to other dysphoria-related conditions? These and similar questions highlight a potential inconsistency in the medical community's approach to BID. For instance, Gender Dysphoria is widely recognised and treated with various interventions, including hormone therapy and gender-affirming surgeries. Similarly, conditions like body dysmorphic disorder, where the primary issue is severe dissatisfaction with one's body, are treated with psychological support and, in some cases, surgery. Given that BID also involves significant psychological distress related to a perceived mismatch between one's physical body and one's sense of Self, it would seem consistent to provide comparable treatment options. This includes not only psychological support but also potential surgical interventions if they are deemed to be in the patient's best interest. The comparison between the desire for amputation in individuals with BID and the wish for SRS in those with Gender Dysphoria has been a topic of long-standing debate (Garcia-Falgueras, 2014; Ostgathe et al., 2014). SRS is currently recognised as an effective treatment for individuals experiencing Gender Dysphoria, and it is similarly endorsed by the World Professional Association for Transgender Health (WPATH) (Coleman et al., 2022; Hage & Karim, 2000). According to WPATH guidelines, SRS plays a crucial role in alleviating the distress caused by Gender Dysphoria. Given the similarities between BID and SRS, one of the most contentious questions today is under what conditions public funding should be considered for amputation as a therapeutic approach for individuals desiring impairment. Given the parallels (although contentious) between requests for amputation and SRS, consistency suggests that what is deemed valid for one should also be considered valid for the other. However, it would be unwise to consider only the similarities without addressing potential objections.

The first objection is that BID differs from Gender Dysphoria's SRS because BID is more akin to cosmetic surgery than to SRS. For many years, BID was strictly associated with paraphilia, mental disorders, or extreme body modification. It was only recently that its medical nature was acknowledged, leading to its inclusion in the ICD-11. While the most visible result of the desire for impairment in BID individuals is a significant alteration of their physiology, which has a cosmetic component, it is now clear that this cosmetic aspect is merely a minor part of a disorder influenced by socio-cultural and biological factors. This understanding alone may be sufficient to refute the argument that BID is purely cosmetic. BID is a fully recognised medical condition and should not be compared to elective cosmetic surgery to justify its exclusion from public funding. Additionally, certain types of cosmetic surgery, such as reduction mammoplasty, otoplasty and gynecomastia surgery for men, are eligible for public funding. In these cases, the primary reason for these procedures is often to improve psychological well-being rather than to restore or enhance physiological function (Almazan et al., 2019; Rasko et al., 2019). Psychological distress can be as debilitating as any physical deficiency. Therefore, the notion that BID is only a cosmetic issue is unfounded. BID is a medical condition, and many cosmetic surgeries are publicly funded even when they do not directly affect physiological well-being or bodily functionality. It is also important to note that cosmetic and clinical procedures are not necessarily two distinct and separate realms; they often intersect, as seen in the case of SRS. In SRS, for example, sexual organs are created with both functional and aesthetic goals. For example, female-to-male transgender individuals undergoing phalloplasty frequently require an erectile prosthetic device to engage in penetrative intercourse, as the constructed organ may not be fully functional on its own. Similarly, male-to-female transgender individuals will not be able to give birth, as the surgical procedure does not include the creation of biological reproductive organs. Following the same reasoning, many European countries have legal requirements for transitioning, including compulsory sterilisation (Dunne, 2017), which clearly and purposely impact a specific type of functionality. Therefore, the argument equating BID to cosmetic surgery lacks a solid foundation and should be dismissed entirely. BID should not be conflated with elective cosmetic procedures due to its recognition as a medical condition with significant psychological and socio-cultural implications. The comparison fails to acknowledge the nuances of BID.

The second objection is that funding surgeries for BID could divert financial resources, time and effort from actual life-saving treatments, raising concerns about scarcity and the fair distribution of healthcare resources (Beauchamp & Childress, 2013, pp. 235–272). This argument presupposes that addressing the desire for impairment in BID is less critical than treating other medical conditions deemed more serious. Consequently, elective amputation is viewed as unnecessary, dangerous and unjustifiable. However, both SRS and BID involve significant dysphoria, which can lead to severe consequences such as self-injury, suicide or, in the case of BID, self-amputation. Additionally, other types of therapy have proven not successful in treating individuals diagnosed with BID. Surgery represents a lower financial burden than years of ongoing therapy or pharmacological treatment, especially when such long-term treatments are covered by statutory health insurance. A relevant comparison can be made with the debate over the medicalisation of Female Genital Mutilation as a harm reduction strategy (Leye et al., 2019; Nabaneh & Muula, 2019; Pearce & Bewley, 2014; Shell-Duncan, 2001). A potential solution to the issue of using public funds for BID-related amputations is to consider allowing individuals to undergo the procedure at their own expense. However, this approach presents two issues.

Firstly, if BID is recognised as a life-threatening condition, the healthcare system should address it with the same urgency and resource allocation as other similar conditions. Denying public funding for BID surgeries while recognising the condition's severity could lead to significant concerns regarding healthcare access. Secondly, limiting surgical intervention to those with the financial means to pay for it raises serious ethical concerns. If BID is life-threatening, restricting treatment based on economic resources is clearly not the way forward. Given these considerations, it is reasonable to argue that individuals with BID should have their surgical requests covered by statutory health insurance plans. Alternatively, one could try to identify morally relevant distinctions between BID and SRS to justify different treatment approaches.

A possible follow-up question to this position is the potential for post-operative complications, which might appear to be more prevalent for BID surgeries than for SRS. While SRS involves the reconstruction of genitalia, BID-related amputation does not involve transforming the limb into something else; instead, the limb is entirely removed (to consider phantom limb pains, which can be to every extent disabling Dijkstra et al., 2002). In other words, to what extent should potential future complications be managed at the expense of the healthcare system? A straightforward answer might be that if public funding is provided for a specific medical treatment, consistency dictates that such funding should also cover any potential future complications arising from that procedure. However, it is important to consider the nuances of each case individually. What seems clear is that limb amputation for individuals with BID is the only currently available procedure that significantly restores or improves their physical and mental health, much like SRS does for transgender individuals (Wierckx et al., 2011).

3.7 Conclusion

In the present chapter, the bioethical principle of justice has been discussed, and potential obstacles to approving surgical amputation through healthcare providers were examined. Remarkably, there seems to be no solid basis for supporting SRS while rejecting BID-related amputations. Both conditions present strong parallels and should be treated equally. This leads to a compelling argument: the desire for amputation among individuals with BID is equivalent to the wish for SRS among those with Gender Dysphoria. Given that SRS is publicly funded, it follows that amputation for BID should also receive public funding. However, it is important to clarify that this section does not advocate for amputation for BID patients. Instead, it aims to shed light on some of the most critical challenges healthcare providers face today when dealing with such requests and some of the gaps in medical knowledge and approach when talking about BID.

References

Allard, J., & Fortin, M.-C. (2017). Organ donation after medical assistance in dying or cessation of life-sustaining treatment requested by conscious patients: The Canadian context. *Journal of Medical Ethics, 43*(9), 601–605. https://doi.org/10.1136/medethics-2016-103460

Almazan, A. N., Boskey, E. R., Labow, B., & Ganor, O. (2019). Insurance policy trends for breast surgery in cisgender women, cisgender men, and transgender men. *Plastic & Reconstructive Surgery, 144*(2), 334e–336e. https://doi.org/10.1097/PRS.0000000000005852

American Psychiatric Association (Ed.). (2013). *Diagnostic and statistical manual of mental disorders: DSM-5* (5th ed.). American Psychiatric Association.

Appelbaum, P. S. (2007). Assessment of patients' competence to consent to treatment. *New England Journal of Medicine, 357*(18), 1834–1840. https://doi.org/10.1056/NEJMcp074045

Appelbaum, P. S., & Grisso, T. (1988). Assessing patients' capacities to consent to treatment. *New England Journal of Medicine, 319*(25), 1635–1638. https://doi.org/10.1056/NEJM198812223192504

Beauchamp, T. L., & Childress, J. F. (2013). *Principles of biomedical ethics* (7th ed.). Oxford University Press.

Beckford-Ball, J. (2000). The amputation of healthy limbs is not an option. *British Journal of Nursing, 9*(4), 188. https://doi.org/10.12968/bjon.2000.9.4.6371

Blom, R. M., Hennekam, R. C., & Denys, D. (2012). Body integrity identity disorder. *PLoS One, 7*(4), e34702. https://doi.org/10.1371/journal.pone.0034702

Blom, R. M., Braam, A. W., de Boer-Kreeft, N., Sonnen, M. P., & a. M. (2014). Body integrity identity disorder, relief after amputation. *Tijdschrift voor Psychiatrie, 56*(1), 54–57.

Chand, N., Subramanya, H., & Rao, G. (2014). Management of patients who refuse blood transfusion. *Indian Journal of Anaesthesia, 58*(5), 658–664. https://doi.org/10.4103/0019-5049.144680

Chung, K. C., Pushman, A. G., & Bellfi, L. T. (2009). A systematic review of ethical principles in the plastic surgery literature. *Plastic and Reconstructive Surgery, 124*(5), 1711–1718. https://doi.org/10.1097/PRS.0b013e3181b98a9f

Cohen, S. (2019). The logic of the interaction between beneficence and respect for autonomy. *Medicine, Health Care and Philosophy, 22*(2), 297–304. https://doi.org/10.1007/s11019-018-9876-4

Coleman, E., Radix, A. E., Bouman, W. P., Brown, G. R., De Vries, A. L. C., Deutsch, M. B., Ettner, R., Fraser, L., Goodman, M., Green, J., Hancock, A. B., Johnson, T. W., Karasic, D. H., Knudson, G. A., Leibowitz, S. F., Meyer-Bahlburg, H. F. L., Monstrey, S. J., Motmans, J., Nahata, L., et al. (2022). Standards of Care for the Health of transgender and gender diverse people, version 8. *International Journal of Transgender Health, 23*(Sup 1), S1–S259. https://doi.org/10.1080/26895269.2022.2100644

Coltheart, M., Langdon, R., & McKay, R. (2007). Schizophrenia and monothematic delusions. *Schizophrenia Bulletin, 33*(3), 642–647. https://doi.org/10.1093/schbul/sbm017

Darby, R. R., & Dickerson, B. C. (2017). Dementia, decision making, and capacity. *Harvard Review of Psychiatry, 25*(6), 270–278. https://doi.org/10.1097/HRP.0000000000000163

Davies, M., Coltheart, M., Langdon, R., & Breen, N. (2001). Monothematic delusions: Towards a two-factor account. *Philosophy, Psychiatry, & Psychology, 8*(2), 133–158. https://doi.org/10.1353/ppp.2001.0007

Dijkstra, P. U., Geertzen, J. H. B., Stewart, R., & Van Der Schans, C. P. (2002). Phantom pain and risk factors: A multivariate analysis. *Journal of Pain and Symptom Management, 24*(6), 578–585. https://doi.org/10.1016/S0885-3924(02)00538-9

Druss, B. G., Marcus, S. C., Olfson, M., & Pincus, H. A. (2002). The most expensive medical conditions in America. *Health Affairs, 21*(4), 105–111. https://doi.org/10.1377/hlthaff.21.4.105

Dudley, R. E. J., John, C. H., Young, A. W., & Over, D. E. (1997). Normal and abnormal reasoning in people with delusions. *British Journal of Clinical Psychology, 36*(2), 243–258. https://doi.org/10.1111/j.2044-8260.1997.tb01410.x

Dunne, P. (2017). Transgender sterilisation requirements in Europe. *Medical Law Review, 25*(4), 554–581. https://doi.org/10.1093/medlaw/fwx028

Engelhardt, H. T. (1974). The disease of masturbation: Values and the concept of disease. *Bulletin of the History of Medicine, 48*(2), 234–248.

Frankfurt, H. G. (1969). Alternate possibilities and moral responsibility. *The Journal of Philosophy, 66*(23), 829–839. https://doi.org/10.2307/2023833

Frankfurt, H. G. (1971). Freedom of the will and the concept of a person. *The Journal of Philosophy, 68*(1), 5–20. https://doi.org/10.2307/2024717

Garcia-Falgueras, A. (2014). Gender dysphoria and body integrity identity disorder: Similarities and differences. *Psychology, 05*(02), 160–165. https://doi.org/10.4236/psych.2014.52025

Gillon, R. (1994). Medical ethics: Four principles plus attention to scope. *BMJ, 309*(6948), 184–188. https://doi.org/10.1136/bmj.309.6948.184

Gillon, R. (2003). Ethics needs principles – Four can encompass the rest – And respect for autonomy should be "first among equals". *Journal of Medical Ethics, 29*(5), 307–312. https://doi.org/10.1136/jme.29.5.307

Gordon, J.-S., Rauprich, O., & Vollmann, J. (2011). Applying the four-principle approach. *Bioethics, 25*(6), 293–300. https://doi.org/10.1111/j.1467-8519.2009.01757.x

Hacking, I. (2006). Making up people. *London Review of Books, 28*(16).

Hacking, I. (2007). Kinds of people: Moving targets: British Academy lecture. In P. J. Marshall (Ed.), *Proceedings of the British Academy, volume 151, 2006 lectures* (1st ed.). British Academy. https://doi.org/10.5871/bacad/9780197264249.001.0001

Hage, J. J., & Karim, R. B. (2000). Ought GIDNOS get nought? Treatment options for nontranssexual gender dysphoria. *Plastic & Reconstructive Surgery, 105*(3), 1222–1227. https://doi.org/10.1097/00006534-200003000-00063

Harris, J. (2003). In praise of unprincipled ethics. *Journal of Medical Ethics, 29*(5), 303–306. https://doi.org/10.1136/jme.29.5.303

Jeong, S., Hwang, Y., Hong, Y., Chu, M., Yun, O., Jeong, Y., Lee, J., & Cho, W. (2020). Withdraw life-sustaining treatment and organ donation. *Transplantation, 104*(S3), S263. https://doi.org/10.1097/01.tp.0000699816.83210.d2

Johnston, J., & Elliott, C. (2002). Healthy limb amputation: Ethical and legal aspects. *Clinical Medicine, 2*(5), 431–435. https://doi.org/10.7861/clinmedicine.2-5-431

Lesieur, O., Mamzer, M.-F., Leloup, M., Gonzalez, F., Herbland, A., Hamon, B., Viallard, M.-L., & Hervé, C. (2013). Eligibility of patients withheld or withdrawn from life-sustaining treatment to organ donation after circulatory arrest death: Epidemiological feasibility study in a French intensive care unit. *Annals of Intensive Care, 3*(1), 36. https://doi.org/10.1186/2110-5820-3-36

Leye, E., Van Eekert, N., Shamu, S., Esho, T., Barrett, H., & ANSER. (2019). Debating medicalization of female genital mutilation/cutting (FGM/C): Learning from (policy) experiences across countries. *Reproductive Health, 16*(158), 1–10. https://doi.org/10.1186/s12978-019-0817-3

Nabaneh, S., & Muula, A. S. (2019). Female genital mutilation/cutting in Africa: A complex legal and ethical landscape. *International Journal of Gynecology & Obstetrics, 145*(2), 253–257. https://doi.org/10.1002/ijgo.12792

Ostgathe, A., Schnell, T., & Kasten, E. (2014). Body integrity identity disorder and gender dysphoria: A pilot study to investigate similarities and differences. *American Journal of Applied Psychology, 3*(6), 138–243. https://doi.org/10.11648/j.ajap.20140306.14

Pearce, A. J., & Bewley, S. (2014). Medicalization of female genital mutilation. Harm reduction or unethical? *Obstetrics, Gynaecology & Reproductive Medicine, 24*(1), 29–30. https://doi.org/10.1016/j.ogrm.2013.11.003

Pellegrino, E. D., & Thomasma, D. C. (1987). The conflict between autonomy and beneficence in medical ethics: Proposal for a resolution. *The Journal of Contemporary Health Law and Policy, 3*, 23–46.

Rasko, Y. M., Rosen, C., Ngaage, L. M., AlFadil, S., Elegbede, A., Ihenatu, C., Nam, A. J., & Slezak, S. (2019). Surgical management of gynecomastia: A review of the current insur-

ance coverage criteria. *Plastic & Reconstructive Surgery, 143*(5), 1361–1368. https://doi.org/10.1097/PRS.0000000000005526
Reich, W. T. (1995). *Encyclopedia of bioethics* (Rev. ed.). Simon & Schuster Macmillan.
Shell-Duncan, B. (2001). The medicalization of female "circumcision": Harm reduction or promotion of a dangerous practice? *Social Science & Medicine, 52*(7), 1013–1028. https://doi.org/10.1016/S0277-9536(00)00208-2
Smith, C. M. (2005). Origin and uses of primum non nocere – Above all, do no harm! *The Journal of Clinical Pharmacology, 45*(4), 371–377. https://doi.org/10.1177/0091270004273680
Suntharalingam, C., Sharples, L., Dudley, C., Bradley, J. A., & Watson, C. J. E. (2009). Time to cardiac death after withdrawal of life-sustaining treatment in potential organ donors. *American Journal of Transplantation, 9*(9), 2157–2165. https://doi.org/10.1111/j.1600-6143.2009.02758.x
Takala, T. (2001). What is wrong with global bioethics? On the limitations of the four principles approach. *Cambridge Quarterly of Healthcare Ethics, 10*(1), 72–77. https://doi.org/10.1017/S0963180101001098
Wierckx, K., Van Caenegem, E., Elaut, E., Dedecker, D., Van De Peer, F., Toye, K., Weyers, S., Hoebeke, P., Monstrey, S., De Cuypere, G., & T'Sjoen, G. (2011). Quality of life and sexual health after sex reassignment surgery in transsexual men. *The Journal of Sexual Medicine, 8*(12), 3379–3388. https://doi.org/10.1111/j.1743-6109.2011.02348.x

Chapter 4
Legitimacy of Bodily Modification Interventstions

4.1 Introduction

Nowadays, altering one's physical appearance is not only possible but often encouraged (Chavoin, 2003; Koblenzer, 2003), which raises questions about potentially non-therapeutic surgical and non-surgical practices aimed not at curing any bodily disease but at enhancing an individual's physical and physiological well-being. Additionally, these medical practices seem to operate mainly within a free-market economy, lacking strict regulatory guidelines. By being a consumer-oriented, heavily advertised entrepreneurial practice, cosmetic surgery is first and foremost (arguably) of elective nature: this raises complex ethical and practical questions. Among these are how to effectively balance the risk-benefit ratio to achieve results that meet the patient's expectations, how to accurately evaluate the patient's comprehension of the inherent risks associated with the procedure and, more broadly, how to assess the surgeon's overall competence in performing such elective interventions. Lastly, is it still possible to maintain a clear difference between therapeutic and elective treatments, which seems to emphasise a dualistic view of mind versus body? Such questions furthermore move the focus to an additional point: the idea of normativity.

> Normative here, in turn, implies a moral component: in traditional practices, the bindingness of activities expresses precepts about how things should or should not be done. (Giddens, 1991, p. 145)

Normativity relates to social norms, specifically what is considered acceptable and encouraged within a given sociocultural context. The relationship between cosmetic surgery and normativity can be interpreted in two seemingly opposing ways. On one hand, cosmetic surgery, with its aim to enhance physical features—sometimes to an extreme degree—can point towards an ideology of individualism or the desire to stand out from the crowd. This is evident in extreme body modification practices, such as oversized breast augmentation or rib removal surgery for aesthetic

L. Loriga, *Body Integrity Dysphoria and the Ethical Dilemma of On-Demand Amputation*, https://doi.org/10.1007/978-3-031-82443-2_4

purposes. On the other hand, the desire to stand out can also be driven by a need to conform to a generally accepted standard of beauty or, more precisely, a standard of normality within a specific sociocultural context. Procedures like rhinoplasty, gynecomastia surgery for men, otoplasty and forehead shaving surgery are often sought not merely to achieve a personal ideal but to align one's appearance with what the majority considers normal and acceptable. In this dual interpretation, cosmetic surgery is both a means of expressing individuality and a way of adhering to societal norms. The idea of beauty falls at the intersection of these two approaches to cosmetic surgery: the wish to stand out from the masses and the longing to conform to an ideal agreed-upon appearance. In this second interpretation, beauty aligns with the concept of normativity. The dilemma, then, is whether a beautiful body should adhere to specific social norms or can be defined more broadly. In other words, should beauty be medicalised and applied through scientific measurements, reiterated on the patient's body in an almost assembly-line fashion? This reflection suggests that a decision should be made about whether the surgeon should accommodate the patient's wishes and, if so, to what degree. Alternatively, the surgeon may steer the patient towards a particular set of guidelines.

Reflecting on beauty standards is particularly relevant in the context of cosmetic surgery due to its apparently predominant elective nature. This reflection is crucial for several reasons, including the distinction between the subjective standards of individual surgeons and the professional standards of the field, as well as the criteria surgeons use to assess a patient's capacity to understand the potential medical complications of the desired surgical procedure. The differentiation between professional and subjective standards can also help to better understand the concept of autonomy. Autonomy is not straightforward to interpret; it can be framed as the capacity of a rational individual to make informed, uncoerced decisions, including those regarding their medical future.

> …at minimum, self-rule that is free from both controlling interference by others and from limitations, such as inadequate understanding, which prevents meaningful choice. The autonomous individual acts freely in accordance with a self-chosen plan, analogous to the way an independent government manages its territories and sets its policies. A person of diminished autonomy, by contrast, is in some respect controlled by others or incapable of deliberating or acting on the basis of his or her desires and plans. For example, prisoners and mentally retarded individuals often have diminished autonomy. (Beauchamp & Childress, 2013, pp. 101–102)

Examining how autonomy is interpreted and assessed within the cosmetic surgery field can provide valuable insights into the inclusion and exclusion criteria that surgeons use to either accept or reject patients seeking elective procedures, which are often not considered life-saving treatments from a legal perspective, such as an appendectomy. Moreover, analysing these criteria enables a deeper examination of the surgeon's perceptions of normality and abnormality. This analysis involves distinguishing between what constitutes a typical patient with a reasonable request and an atypical patient with an unusual request. Such an investigation can shed light on the intricate nature of Body Integrity Dysphoria (BID) and the degree to which individuals can make autonomous decisions regarding their bodies. To delve deeper into this subject, it is crucial to base the analysis on primary data sources.

4.2 Background

This chapter will focus on two pivotal topics that emerged during the fieldwork research phase: the concept of body harmony and the concept of normativity. Over 36 months, extensive primary data was gathered through direct engagement with various medical stakeholders. A total of approximately one hundred cosmetic surgery clinics were contacted using a combination of in-person visits, e-mails and phone calls, with many instances involving all three methods. In addition to reaching out to cosmetic surgery clinics, the same engagement strategies were employed to connect with both profit and non-profit organisations dealing with body-related matters, as well as Clinical Psychologists and Psychiatrists. The data collection process involved conducting semi-structured interviews, which varied in length from a minimum of 14 min to a maximum of 47 min, with an average duration of 33 min per interview. All participants expressed their willingness to engage in follow-up interviews. Prior to the commencement of each interview, an information sheet and consent form were provided, ensuring that both the interviewer and interviewee were fully informed and in agreement. In cases where the interview was recorded, explicit consent was obtained from the participants.

This chapter will analyse and interpret the primary data collected during this extensive research, which provides valuable insights into the perceptions and practices related to body harmony and normativity within the field of cosmetic surgery. For a detailed overview of the quantitative data related to this project, refer to Loriga (Loriga, 2024).

4.3 Themes of Discussion

Although body harmony and normativity might appear synonymous within the context of cosmetic surgery, they emerged, in fact, as distinct concepts. Body harmony refers to the balanced metrics and proportions of the body. At the same time, normativity in this context pertains more to the patient's expectations and motivations for undergoing surgery than to their physical appearance. Both concepts can independently or jointly provide valid reasons for seeking surgery. At this intersection, the surgeon assesses the patient's medical capacity during the informed consent process. It is here that the distinction between the surgeon's professional standard and subjective standard becomes evident. The professional standard is grounded in medical ethics and best practices. In contrast, the subjective standard is influenced by a range of factors, including the surgeon's gender, age, sociocultural background, personal experiences and political beliefs.

4.4 Normativity

A good starting point is the concept of normativity as it applies to cosmetic surgery practices. From here, it becomes crucial to examine what is considered normal and abnormal within this context. Judgements of normalcy can vary significantly between doctors and patients for various reasons. For instance, a prominent nose might be viewed as an abnormality in some contexts but as perfectly ordinary in others. Similarly, invasive cosmetic procedures like extreme breast augmentation can be perceived differently depending on individual perspectives and cultural standards. During numerous interviews with cosmetic and reconstructive surgeons, it became evident that they consider two distinct parameters when evaluating the legitimacy of a patient's request for surgery: motivations and expectations. Motivations refer to the reasons behind a patient's desire for surgery, while expectations pertain to the anticipated outcomes of the procedure. Both parameters are essential in the surgeon's assessment of the patient's request and are used to evaluate the patient's medical decision-making capacity or autonomy. Three potential postulates might disqualify patients from being considered for surgical intervention, while only one would allow them to proceed with the desired medical intervention.

1. The patient's motivation for surgery is not a sufficient reason to counterbalance the post-operative risks.
2. The patient's motivation for surgery is inherently invalid, meaning that such motivations undermine the individual's medical decisional capacity.
3. The patient's motivation for surgery is valid, but the expected outcomes cannot be realistically achieved.
4. The patient's motivation for surgery is sufficient, and the expected outcomes are achievable.

From this brief schematisation, one can observe a causal relationship between motivation, expectation and the surgeon's evaluation. However, despite this straightforward framework, it is only in the last postulate that the concept of normalcy can be reframed in favour of the patient. In other words, the correlation between motivation and expectation shifts the evaluation of normal and abnormal from social standards to the subjective standard of the doctor-patient relationship. The following pages will illustrate this latter point. However, before doing that, it is necessary to briefly explain the first three points.

4.4.1 Postulate One

The patient's motivation for undergoing surgery, no matter how compelling, does not justify proceeding if the post-operative risks outweigh the potential benefits. In other words, if the risk-benefit ratio is disproportionately skewed towards risk, the patient should be deemed ineligible for the desired procedure. It is important to recognise that the well-being of both the patient and the surgeon is at risk in cosmetic surgery, particularly given its elective nature. As cosmetic surgery is inherently an entrepreneurial practice, agreeing to perform the procedure not only involves medical judgement but also carries significant liability for both parties involved.

For instance, consider a patient who desires a high-risk cosmetic procedure, such as an extensive body contouring surgery, following significant weight loss. While the patient may be highly motivated to achieve a certain aesthetic outcome, the potential complications—such as poor wound healing, infection or even life-threatening issues like deep vein thrombosis—might outweigh the perceived benefits. In this scenario, even though the patient is eager and determined, the surgeon has a professional and ethical obligation to refuse the surgery if the risks are deemed too great. By doing so, the surgeon is safeguarding both the patient's health and their professional integrity, avoiding a situation where the liabilities could be substantial for both parties.

4.4.2 Postulate Two

The patient's motivation for undergoing surgery may be considered inherently invalid if, in the surgeon's professional judgement, such motivations compromise the patient's capacity to make sound medical decisions. This involves making a critical evaluative judgement regarding the patient's autonomy. For example, the motivation and desire to pursue surgery must genuinely originate from the patient. Requests for medical-surgical procedures, particularly those that are elective and not necessary for addressing life-threatening conditions, inherently carry risks; therefore, they must be initiated by the patient themselves. If this criterion is not satisfied, the patient should be disqualified from receiving the procedure. In discussions with cosmetic surgery professionals, there was a categorical and unanimous agreement on this point. However, they emphasised that the origin of the motivation and desire to undergo surgery cannot be the sole criterion in determining its legitimacy; it serves merely as the initial step in a broader evaluation. For instance, consider a patient who expresses a desire for rhinoplasty. Suppose the motivation for this surgery stems from external pressures—such as a partner's insistence or societal expectations—rather than a personal and autonomous desire. In that case, the surgeon might view this as undermining the patient's decision-making capacity. However, the source of the desire alone is not enough to assess its legitimacy. During interviews, two additional sub-criteria were identified: the duration of the desire and the degree of compulsion behind it. For example, suppose a patient had the desire for rhinoplasty for several years, and this desire is consistent and not impulsive. In that case, the longevity of the wish may support its legitimacy. Conversely, if the desire is driven by an obsessive compulsion rather than a reasoned and well-considered decision, it may signal that the patient's autonomy is compromised. Thus, the surgeon must evaluate both the persistence and the psychological underpinnings of the patient's motivation to ensure that the decision to proceed with surgery is both autonomous and informed.

4.4.2.1 Desire Compulsion

Desire compulsion is compared to a craving that negatively impacts the patient's medical decision-making capacity. This craving undermines the informed consent process within the doctor-patient relationship, which is crucial not only for the patient but also for the surgeon to feel confident that the patient fully understands the procedure. Such a compulsive craving disqualifies the individual from being considered a rational, autonomous agent capable of comprehending the information regarding their medical future. The criterion of desire compulsion is, therefore, significant for several reasons, primarily the surgeon's assessment of the patient's rationality. This assessment includes evaluating the patient's understanding of not only what they are requesting but also potential future medical complications.

4.4.2.2 Desire Duration

The desire duration sub-criterion is the minimum time a specific desire must persist to be considered legitimate. Desire duration helps assess the impulsiveness of the patient's request, which might be based on a temporary state of dissatisfaction due to everyday life occurrences. This criterion is a further step from desire compulsion. While desire compulsion allows the surgeon to evaluate the individual's ability to reflect clearly on their motivations, desire duration filters out temporary factors that might cloud the individual's judgement. These two levels of analysis—desire compulsion and desire duration—are therefore complementary. Together, they provide a comprehensive assessment of the patient's capacity to make a well-considered and informed decision about undergoing cosmetic surgery.

4.4.3 Postulate Three

The patient's motivation for surgery may be valid, but the expectations they have for the outcomes cannot be realistically achieved. Within this category, two extensive thematic areas emerged among all respondents: Body Dysmorphic Disorder (BDD) and the impact of social media. These criteria have two notable peculiarities: their flexibility, referring to the degree to which they are present in the patient, and their temporality, meaning their relevance to a specific point in time or era.

4.4.3.1 Body Dysmorphic Disorder

Individuals with BDD experience a persistent preoccupation with minor perceived physical flaws, often accompanied by ideas of reference—the fear or delusion of being discussed or judged because of those flaws (World Health Organization, 2019, 6B21). BDD sufferers frequently seek cosmetic surgery to alleviate their discomfort (De Brito et al., 2013; Felix et al., 2014). The condition of BDD and its impact on an individual's medical decision-making capacity can illuminate the relationship between suffering and patient autonomy.

To explore this correlation, one can use the concept of the window of tolerance, first introduced by Siegel (2020) and clearly described by a respondent from the Clinical Psychology group. The window of tolerance is a hypothetical boundary, depicted as the squared frame of a window, within which an individual can experience various levels and types of discomfort while maintaining rationality and self-control. This boundary represents the emotional space familiar to the individual, allowing them to navigate discomfort without losing autonomy. Within the window of tolerance, individuals can endure different forms and intensities of suffering while still being in control of their emotions and decisions. However, once they move outside this boundary, the emotional aspect overwhelms the rational one, leading to a loss of autonomy.

This depiction is beneficial in illustrating that autonomy is a matter of degree rather than a black-and-white concept. The exact positioning of these fictitious perimeters for each individual and the feasibility of recognising them from an external perspective, particularly in the context of elective medical interventions, is a complex issue. This topic demands further discussion and goes beyond the scope of the present chapter.

4.4.3.2 Social Media

Regarding social media, it is undeniable that cosmetic surgery requests have surged significantly in recent years. Several factors contribute to this trend, including consumer pressure to conform to prevailing beauty standards (Sarwer et al., 2003) and heightened media exposure (Brown et al., 2007; Delinsky, 2005; Markey & Markey, 2009; Swami et al., 2008). Today's media-driven culture perpetuates a beauty bias that fosters increased bodily awareness, characteristic of the image-based culture of postmodernity. This beauty bias compels individuals to strive for often unattainable physical ideals, leading to a growing demand for cosmetic procedures. The pervasive influence of social media platforms exacerbates this phenomenon, as users are continually exposed to edited and curated images that set unrealistic standards of beauty. Moreover, the constant comparison facilitated by social media can significantly impact self-esteem and body image, driving more individuals to seek cosmetic surgery as a means of achieving perceived societal ideals.

4.4.4 *Postulate Four*

The patient's motivation for surgery is valid, and their expectations from the procedure can be realistically achieved. This postulate differs from the other three for two primary reasons. Firstly, it is the only one that permits the individual to proceed with the surgery or continue with the assessment process. Here, the patient's motivations and expectations are aligned, creating a clear pathway to the desired medical intervention. Within the framework of normalcy, this alignment is crucial: normal motivations lead to normal expectations, thereby legitimising the patient's desire. This one-to-one correspondence is essential for achieving the sought results. An abnormal motivation cannot justify normal expectations, and likewise, abnormal expectations cannot be justified by normal motivations.

4.4.4.1 The Legitimacy of Abnormality

The fourth postulate therefore establishes a one-to-one relationship between motivations and expectations, suggesting that this direct correlation justifies the patient's request. This relationship can be either normal or abnormal. This distinction is crucial; the one-to-one correlation between motivations and expectations maintains its validity even when both are abnormal, as long as the end goal is recognised as rational. In other words, if the patient's motivations and expectations are both abnormal but still lead to a rational and attainable outcome, the request can be considered legitimate. This implies that even an unconventional or extraordinary request, such as a significantly exaggerated breast augmentation procedure, can be deemed valid if the individual is able to clearly and convincingly articulate their reasons for pursuing such a surgery. The key factor is not necessarily the nature of the request but rather the patient's ability to demonstrate a well-considered and genuine motivation behind their decision. For instance, during interviews with cosmetic surgery professionals, several respondents provided practical examples to illustrate the legitimacy of such requests. One common scenario involved individuals working in the entertainment industry, where physical appearance plays a crucial role in professional success. For example, performers such as dancers, adult film actresses or others in show business might request breast augmentation to achieve a look that aligns with industry standards or enhances their marketability within their specific field. In these cases, the request for a larger-than-average breast size might initially seem extreme or out of the ordinary. However, when these individuals can explain that augmentation is necessary to maintain or advance their careers, the request gains legitimacy in the eyes of the surgeon. In these cases, the rationale behind the procedure is linked to professional requirements or personal branding, where physical appearance plays a significant role in career success. Thus, even if the request appears extreme, it can be deemed legitimate if the individual can provide a rational and coherent explanation for their decision. This perspective recognises that individuals have the right to make informed decisions about their bodies, even if those decisions involve a degree of risk. As long as the reasons are well-articulated and grounded in a rational perspective, the request can be considered valid. This concept of legitimising out-of-the-ordinary motivations and expectations is particularly relevant for BID and will be revisited in due course.

4.5 Body Harmony

Let's now shift the focus to body harmony, the harmonious balance of bodily metrics and proportions. This concept is inherently intertwined with the idea of naturalness, necessitating a distinction between a natural and unnatural appearance.

> …Proportions should be balanced according to the height, weight, and shape of the lower and upper body parts. The waist and breasts should be well-balanced—neither too big nor too small. The nose should correspond to the width of the face, maintaining harmonic proportions. I explain it to patients like this: If a part of your body is 'screaming' for attention, it's disharmonic. After surgery, that part of the body should no longer stand out; everything should be done proportionally and harmoniously... *Michal, M.D. Cosmetic & Reconstructive Surgery*

From this brief example, which is representative of many similar cases, one can deduce a strong correlation between the concepts of harmony and beauty. The alignment of these two ideas suggests that what is often perceived as beautiful is intrinsically linked to a sense of harmony or balance, whether in physical appearance, artistic expression or natural forms. The implication is that beauty is deeply rooted in the harmonious arrangement of elements, be they features of the human body or aspects of a broader composition. In this interpretation, beauty becomes a universal phenomenon, characterised by ideal proportions expressed in geometrical terms, often referred to as the golden ratio—or divine proportion—which has long been associated with anatomical beauty (Persaud-Sharma & O'Leary, 2020; Rupesh et al., 2014). This golden ratio is seen as an intrinsic standard of beauty, harmonising various elements of the body to achieve an aesthetically pleasing result. The notion of beauty, which correlates closely with the idea of harmony, was consistently identified as the standard to strive for in all the interviews conducted. Surgeons aim to achieve this ideal of beauty by adhering to principles of proportion and balance, ensuring that their work enhances the natural symmetry and attractiveness of the patient's features. Therefore, this standard, directly related to an individual's body shape, is considered the healthy or natural baseline from which all variations are measured. Consequently, any request from a patient that deviates from these ideal proportions of natural body shape is a priori considered potentially harmful and subject to much stricter scrutiny by the surgeon. As a result, the patient's autonomy is questioned if they express a willingness to conform to something that does not align with these standard ideals. Surgeons may view such requests as indicative of a potential misunderstanding or disregard for what is considered healthy and natural, thereby necessitating a deeper evaluation of the patient's motivations and expectations. From here, it naturally follows that an amputation request, no matter how rational it might seem, inherently goes against the established standard of body harmony and natural proportions. Such a request deviates significantly from what is considered healthy and ideal, prompting heightened scrutiny and questioning of the patient's autonomy and motivations.

4.6 Conclusion

The chapter delved into two significant topics that emerged during the interview process: normativity and body harmony. These themes were central to the discussions, reflecting the concerns and considerations of professionals within the field of cosmetic surgery. By examining these topics, the chapter sheds light on how these concepts influence decision-making processes and shape perceptions within medical practice. These concepts, whether considered interconnected or standalone, were employed to delve into the notion of medical decision-making capacity—and consequently, autonomy—from the medical standpoint. The aim was to identify the main criteria that would qualify or disqualify a patient from being admitted for the desired surgical procedure. In the discussion of normativity, the chapter explored the patient's motivations and expectations from the perspective of the physician, emphasising their importance in evaluating the legitimacy of the patient's desire for surgery. Through this examination, four distinct postulates were formulated by combining various motivations and expectations. Among these, it was determined that only one—the one-to-one formula—would allow the patient to pass the initial psychological assessment. This formula suggests that when a patient's motivations and expectations align closely and consistently, the desire for surgery can be deemed legitimate according to the surgeon's subjective standard, even if these desires deviate from broader societal norms. This assessment highlights that a patient with consistent motivations and expectations may still be considered a suitable candidate for surgery, provided their reasons are coherent and sincere. The surgeon's role, therefore, involves not just evaluating the technical aspects of the procedure but also making nuanced judgements about the patient's psychological readiness and the authenticity of their desire, even in cases where societal standards might typically view such desires as unconventional.

In the final part of the chapter, the concept of body harmony was presented. A comparison was highlighted between harmony, the idea of beauty, health and nature, which can be summarised in the following sequence: harmony is to beauty as disharmony is to ugly, and beauty is to health as ugly is to harm. It necessitates that health is to nature as harm is to artificial which is per se a paradox because cosmetic surgery deals with the artificial rather than with the natural and despite that, it professes the natural, harmonic, healthy look.

References

Beauchamp, T. L., & Childress, J. F. (2013). *Principles of biomedical ethics* (7th ed.). Oxford University Press.

Brown, A., Furnham, A., Glanville, L., & Swami, V. (2007). Factors that affect the likelihood of undergoing cosmetic surgery. *Aesthetic Surgery Journal, 27*(5), 501–508. https://doi.org/10.1016/j.asj.2007.06.004

Chavoin, J. P. (2003). Chirurgie esthétique et éthique. *Annales de Chirurgie Plastique Esthétique, 48*(5), 273–278. https://doi.org/10.1016/S0294-1260(03)00079-7

De Brito, M. J. A., Nahas, F. X., Ortega, N. R. S., Cordás, T. A., Dini, G. M., Neto, M. S., & Ferreira, L. M. (2013). Support system for decision making in the identification of risk for body dysmorphic disorder: A fuzzy model. *International Journal of Medical Informatics, 82*(9), 844–853. https://doi.org/10.1016/j.ijmedinf.2013.04.007

Delinsky, S. S. (2005). Cosmetic surgery: A common and accepted form of self-improvement? 1. *Journal of Applied Social Psychology, 35*(10), 2012–2028. https://doi.org/10.1111/j.1559-1816.2005.tb02207.x

Felix, G. A. A., De Brito, M. J. A., Nahas, F. X., Tavares, H., Cordás, T. A., Dini, G. M., & Ferreira, L. M. (2014). Patients with mild to moderate body dysmorphic disorder may benefit from rhinoplasty. *Journal of Plastic, Reconstructive & Aesthetic Surgery, 67*(5), 646–654. https://doi.org/10.1016/j.bjps.2014.01.002

Giddens, A. (1991). The trajectory of the self. In *Modernity and self-identity: Self and society in the late modern age* (pp. 70–109). Stanford University Press.

Koblenzer, C. S. (2003). Psychosocial aspects of beauty: How and why to look good. *Clinics in Dermatology, 21*(6), 473–475. https://doi.org/10.1016/j.clindermatol.2003.11.010

Loriga, L. (2024). The right to choose: A comparative analysis of patient autonomy and body integrity dysphoria among Czech healthcare professionals. *Ethics & Bioethics, 14*(1–2), 41–60. https://doi.org/10.2478/ebce-2024-0001

Markey, C. N., & Markey, P. M. (2009). Correlates of young women's interest in obtaining cosmetic surgery. *Sex Roles, 61*(3), 158–166. https://doi.org/10.1007/s11199-009-9625-5

Persaud-Sharma, D., & O'Leary, J. P. (2020). Fibonacci series, golden proportions, and the human biology. *Austin Journal of Surgery, 2*(5), 1066.

Rupesh, S., Rakesh, S., Winnier, J., Kaimal, A., John, A., Prasannan, M., & Jeyaprakash, V. (2014). The role of divine proportion in perception of beauty: A cross sectional study. *Amrita Journal of Medicine, 10*(1), 1–44.

Sarwer, D. B., Magee, L., & Clark, V. (2003). Physical appearance and cosmetic medical treatments: Physiological and socio-cultural influences. *Journal of Cosmetic Dermatology, 2*(1), 29–39. https://doi.org/10.1111/j.1473-2130.2003.00003.x

Siegel, D. J. (2020). *The developing mind: How relationships and the brain interact to shape who we are* (3rd ed.). Guilford Press.

Swami, V., Arteche, A., Chamorro-Premuzic, T., Furnham, A., Stieger, S., Haubner, T., & Voracek, M. (2008). Looking good: Factors affecting the likelihood of having cosmetic surgery. *European Journal of Plastic Surgery, 30*(5), 211–218. https://doi.org/10.1007/s00238-007-0185-z

World Health Organization. (2019). *International statistical classification of diseases and related health problems* (11th ed.). Retrieved January 6, 2024, from https://icd.who.int/en

Chapter 5
Pre-operative Risk Assessment and Post-operative Risk Behaviours

5.1 Introduction

Competency determination is "a determination of a particular person's capacity to perform a particular decision-making task at a particular time and under specific conditions" (Buchanan & Brock, 1986, p. 23). The concept of competency determination was a consistent theme throughout the entire interview process, sometimes addressed implicitly and at other times explicitly. However, respondents focused on two specific aspects when discussing a patient's fitness to undergo any medical therapy: pre-operative risk assessment and post-operative risk behaviour. Pre-operative risk assessment encompasses all practices and evaluations conducted before surgery to ensure patient safety and understanding. These practices include medical consultations, diagnostic tests and detailed discussions about the procedure's risks and benefits. The goal is to confirm that the patient is fully informed and medically fit for surgery, thereby minimising potential complications. Post-operative risk behaviour, on the other hand, involves monitoring, evaluating and predicting the patient's actions and habits after the surgery, which could jeopardise the procedure's success. This includes assessing whether the patient adheres to post-operative care instructions, maintains proper hygiene, avoids risky activities and follows prescribed medication regimens.

When discussing post-operative risk behaviour, a thought-provoking and intentionally provocative comparison with Body Integrity Dysphoria (BID) appeared: gastric banding and sleeve gastrectomy for obese patients. This topic emerged unexpectedly during research with Clinical Psychologists, proving to be particularly pertinent to the ongoing debate. The patient's medical history, in these cases, offer valuable insights into predicting future behaviours, addressing a common criticism of BID: How can one be certain that an individual fully comprehends the long-term consequences of their decision and will not jeopardise their health or squander public funds allocated for the procedure? These cases provide a unique perspective on

L. Loriga, *Body Integrity Dysphoria and the Ethical Dilemma of On-Demand Amputation*, https://doi.org/10.1007/978-3-031-82443-2_5

the individual's understanding and commitment to the outcomes of their medical decisions. The scope of such comparison, which will be expanded in the following sections, is to identify which elements are considered when evaluating an individual's behavioural pattern to assess fitness for surgery.

A final remark before proceeding is necessary. Despite the apparent division of focus, distinguishing between pre-operative risk assessment and post-operative risk behaviour is only partially accurate. These two aspects are interconnected and should be considered integral components of a comprehensive evaluation of the patient's overall competency and readiness for medical intervention.

5.2 Pre-operative Risk Assessment

Pre-operative risk assessment encompasses all practices undertaken by the patient, following the physician's advice, aimed at reducing risks during invasive and non-invasive medical therapies. These practices range from general consultations for less urgent cases to more complex analyses, such as an echocardiogram before surgery. Within the context of competency determination, these elements provide valuable insights into how the medical field frames and assesses patient autonomy. These assessments help to ensure that the patient is fully informed and capable of understanding the implications of the procedure, thus safeguarding their autonomy while also addressing potential risks. By examining the various factors involved in pre-operative risk assessment, one can appreciate the comprehensive approach taken by healthcare professionals to evaluate patient readiness and competence for medical interventions. What is asked of the surgeon in the case of elective practices is to cast a judgment of competency. Singularly enough, such dynamics often create a power struggle between doctor and patient where the latter needs to present one's motivations and expectations in a manner convincing enough for the doctor to accept such request, while the first needs to use such patient's demands to infer the much broader context of patient's competency and autonomy. Another critical element is the type of capacity involved, which must ensure that meaningful choices are made. In the medical context, medical decision-making capacity refers to the ability to make voluntary, uncoerced and well-considered decisions about one's medical future while understanding the potential consequences. This term is distinct from legal capacity or competency. For clarity, medical decision-making capacity is strictly confined to the clinical context. Competency, on the other hand, is a legal determination that applies more broadly to the individual.

As defined in the preceding sections, the medical decision-making capacity of a patient within an institutionalised medical context is directly related to the extent to which their medical condition is recognised as legitimate by the dominant medical discourse. In other words, a patient reporting a condition that is medically recognised—such as those listed in the ICD and DSM—is far more likely to have their narrative accepted as legitimate compared to a patient reporting a condition that is not medically recognised. Interestingly, a clear correlation emerged from many of the respondents between the patient's medical decision-making capacity and their adherence to medical advice for surgery. This trend suggests that patients who go against the medical advice of their physician are far more likely to be judged as lacking one or more core elements of medical decision-making capacity, particularly regarding their understanding of the potential future consequences of their medical decisions.

In essence, patients whose conditions are recognised and who follow medical advice are seen as more competent in making informed decisions about their health. Conversely, those who defy medical recommendations are perceived as having diminished capacity to fully grasp the implications of their choices. This underlines an inherent paternalistic stance. Patients may be able to make confident decisions about their health but simultaneously unable to make others, depending on the risks, benefits and complexity involved. An example of this dynamic within cosmetic surgery is rhinoplasty, a topic that emerged throughout the interview phase. Although rhinoplasty is one of the most common cosmetic surgery procedures, intervening on the nose completely alters the facial physiognomy of the individual, so it is essential to perform a proper nasofacial analysis before indulging in the surgical procedure. Interestingly, most of the Cosmetic and Reconstructive Surgery group reported being more prone to perform covert invasive surgery than standard rhinoplasty. In other words, an invasive cosmetic surgical procedure which does not alter the apparent physiognomy of the individual—as might be the case for body lift—is deemed to require a lesser degree of autonomy than less invasive surgical procedures that alter the physiognomy of the individual. A fascinating phenomenon arose when such reasoning was applied and proposed to the BID case. To the question of whether it might be more acceptable to remove a non-vital inner organ—assuming the BID desire, as it is understood today, would be re-directed, hypothetically towards that specific end—compared to a limb, the group of the respondents split between those who were more prone to accept the idea of the removal of a non-vital inner organ rather than a limb and those who rejected both instances similarly. No clear justification, however, was provided for either of such decisions besides the idea of functionality and risk-reward ratio.

> …I came across this kind of situation before because I was working in a muscular surgery department many years ago, and there was a lady who had some infected and chronic wounds on her leg. She preferred amputation over treatment every couple of weeks. Of course, we couldn't... we didn't want to... for her, it was a more effective treatment than just going through the psychological stress. But still, it's hard to understand... I wouldn't treat that because it's difficult to draw the line... I think the main issue is maintaining a functional body, which, if you consider amputating limbs or other parts of the body, wouldn't be possible… *Daniel, M.D. Cosmetic & Reconstructive Surgery*

The above excerpt concludes that despite limb amputation being recognised by the physician as a more effective treatment from the patient's side than weekly medications, medical advice is prioritised over the patient's choice. Similarly, limb amputation is a "…hard to draw the line…" procedure, which implies the slippery-slope argument. The same thing, however, does not necessarily apply to non-vital inner organs such as the appendix (Bijnen et al., 2003; Walker et al., 1995), tonsils and even ribs (Verdugo, 2022), which might be removed nowadays for preventive, therapeutic or even cosmetic purposes. A possible argument for such removal procedures might be—arguably— that neither appendix, tonsils, nor ribs are elements of the human body that, if removed, would impact its functionality as is the case for an arm or a leg. On the other hand, it is unfeasible to apply the same concept of functionality to the limbs of individuals recognised as having BID as one would to the limbs of those who do not have BID.

The idea of functionality, therefore, is not an objective and universal term despite often being recognised as such. The patient's wish to go against the medical advice is not per se an element that hints towards an impaired medical decision-making capacity, assuming the patient comprehends its core elements—implicitly or not. A good breakdown of what such elements might be recognising is medical decision-making capacity as being compounded by the ability to communicate a choice (i.e. the preference of having a limb removed rather than going through weekly painful medical procedures), a proper understanding of relevant information (i.e. which equates to the concepts fulfilled through the informed consent process), the ability to appreciate the situation and its consequences (in other words the implications of such specific choice) and finally the capacity to process and analyse information logically, in other words, the ability to use rational thinking to weigh the potential benefits against the risks of different treatment (Appelbaum & Grisso, 1988, p. 1636). Finally, when the respondents were introduced to the idea of a potential wish of impairment affecting cognitive faculties—even partially, for example, hearing loss in the left ear or sight from one eye—the respondent group's response was a unanimous rejection.

5.3 Post-operative Risk Behaviour

Within this context, post-operative risk behaviour is the likelihood of the patient engaging in behaviours that would hinder the recovery from a given surgical procedure. Similarly, post-operative risk behaviours are also the likelihood of maintaining a particular behavioural pattern, which would significantly increase the risk of serious complications post-surgery. Lastly, post-operative risk behaviours can also be applied to all those situations where there is a concern that the finances released by a third party—such as a funding body—might not have the wished outcomes, not because of the inefficacy of the medical procedure but because of the reckless behaviour of the patient.

> …I think that looking into the patient's history is probably the best approach; I need to predict their ability to handle certain situations in the future. For example, if someone has had frequent issues with impulsivity, changing jobs, changing partners, driving under the influence, and so on, those are significant red flags. I would consider the patient's opinion and their environment. The inability to control oneself after surgery could have serious medical consequences... *Kamila, Clinical Psychology & Psychotherapy*

A few elements are identifiable from this and similar snippets: the need to have a longitudinal overview of the patient's behavioural pattern, the patient's motivation for undergoing such a procedure and the patient's sociocultural background. The assessment is only partially the result of a clear and recorded patient medical history. The other half is based on the psychological assessment done by the entity in charge, such as Clinical Psychologists.

> ...There is not truly a scale to measure emotional regulation skills directly, so we are utilising proxy methods such as the Rorschach test. Additionally, we are employing complex personality questionnaires and conducting clinical interviews. These are the primary methods we are using. There are also some recommendations, particularly regarding impulsivity as a red flag, as well as a lack of cognitive skills, which could indicate conditions such as mental retardation or dementia. These conditions would undoubtedly serve as exclusion criteria... *Kamila, Clinical Psychology & Psychotherapy*

Despite that, the Clinical Psychologist's assessment also entails a certain degree of guesswork for the simple fact that inferring how a patient will behave is simply impossible.

> ...We can never fully predict an individual's psychological functioning because it is constantly influenced by their environment. I could conduct an assessment today, but a traumatic event occurring subsequently could drastically alter that person's emotional regulation skills.... *Kamila, Clinical Psychology & Psychotherapy*

Such a statement is exemplary as it conveys the idea of the incognito present in every medical intervention due to the patient's behaviour. Similarly, from such incognito, it is possible to draw some questions such as how much information would the patient need in order to make a meaningful choice about their medical future, how much information would the physician need in order to make sure the patient will not engage in risky behaviour. Additionally, there is the question of how long in the future may the risky behaviour manifest before being considered part of regular life occurrences. Although it is impossible to answer all these questions with accuracy, the patient's medical history and psychological assessment should be satisfactory enough to dissipate any reasonable doubt. Anything more than that would be pure guesswork. The idea of post-operative risk behaviour is particularly relevant for the BID as it draws on some of its main critiques (some of them already addressed in the previous chapters). Some of such critiques gravitate around two reasonings, both of which fall within the spectrum of post-operative risk behaviour.

The first critique is whether the BID individual would eventually regret the amputation once performed. Even though such a question has already been addressed, it is interesting now to expand it further. Stating possible regret after amputation for a BID patient may have the same value as stating possible regret for any irreversible medical procedure. This is, moreover, true if, as previously assessed, amputation is recognised as a non-elective medical intervention. A convincing medical history, which does not present any alarming element, compounded by a psychological assessment aimed at excluding the patient's emotional instability, might be considered to every extent enough to dissipate any reasonable doubt about the patient's situation. Bargaining to obtain an assurance more than this would be unrealistic.

The second critique concerns whether it would be better to postpone the amputation in light of a better understanding of the condition and the putative discovery of non-surgical therapy. Such a point might be considered as frail as the previous one because, ideally, any current medical therapy will eventually change in the long run. Therefore, it is not whether there will be alternatives to amputation in the long run, but rather whether it would be morally and ethically acceptable to deny immediate relief to BID individuals in light of a possible and unforeseen new medical therapy that might not happen at all. In other words, is it better to deny the amputation request for an individual recognised as falling within the BID spectrum based solely on the fact that a new therapy might be available in the future?

When such a variable was introduced to the respondent groups, a relatively significant but not unexpected differentiation started to appear between surgeons, clinical psychologists and psychiatrists, with the former being more inclined to wait for a better alternative based on the hypothetical holiness of the incorruptible body and its functionality[1]. At the same time, the latter were more prone to consider the individual's suffering and the possibility of immediate relief at the expense of bodily integrity.

The following section will consider one particularly insightful case of post-operative risk behaviour assessment: Gastric Banding or Sleeve Gastrectomy for obese patients. Even though little to no parallels exist between BID and obesity, the following case is particularly interesting because it considers two elements previously attributed to BID, namely the impulsivity of the amputation request and the individual's emotional stability.

5.4 Gastric Banding and Sleeve Gastrectomy: A Comparison with BID

As noted in the previous sections, several elements emerged from the post-operative risk behaviour assessment. However, one result is fascinating: understanding the mechanism put in place today to grant access to surgery and funding, which is the case with Gastric Banding or Sleeve Gastrectomy[2] for obese patients. Obesity is an excellent example to analyse post-operative risk behaviours, assuming its occurrence is due to a compulsive eating disorder and not biological imbalances.

---

[1] Note, however, that the responses from the surgeon groups were also not unanimous.

[2] While gastric banding and sleeve gastrectomy are distinct surgical procedures, their differences are not pertinent to the present discussion. What is relevant is that both procedures are subject to evaluation and recommendation by clinical professionals, particularly considering the significance of post-operative risk behaviours. These behaviours, often influenced by obesity-related factors and unregulated life choices, are critical in assessing the long-term success of the surgeries.

Obesity is defined as a "...excessive adiposity that can impair health. It is, in most cases, a multifactorial disease due to obesogenic environments, psycho-social factors and genetic variants" (World Health Organization, 2019, 5B81) and "...results from the long-term excess of energy intake relative to energy expenditure..."caused by"...a range of genetic, physiological, behavioural, and environmental factors..." (American Psychiatric Association, 2013, p. 239). The relevance of obesity in the debate on post-operative risk behaviours emerges on two premises: The assessment of the individual to not endanger oneself after surgery, which connects directly with the second point, which is the request for such procedures to be funded by the Statutory Health Care System or various insurance funds.

> ...The main criterion is whether the patient is suffering from it and has the motivation to do something about it. This is when they may be suitable for psychotherapy and, subsequently, surgery. There are many people who suffer from obesity but are not willing to take responsibility or work on changing it... *Kamila, Clinical Psychology & Psychotherapy*

Such an excerpt highlights the individual's awareness of their current condition and their willingness to improve. Following such reasoning, an individual who experiences a state of suffering but does not act on it to remove it or improve one's condition is deemed not in the right state of mind to undergo any medical procedure to relieve such suffering, especially a surgical procedure. This connects back directly with the BID topic by bringing to the foreground once again the question of whether it would be morally and ethically acceptable to refuse an individual who recognised and endured sufferance for an extended period (reminding that the BID desire is considered to first manifest itself during late childhood and early adolescence) and now asking medical aid to relieve such situation.

> ...I conduct examinations for individuals considering surgery, which is a significant intervention in the body and must be thoroughly considered. It is crucial to ensure that the person fully understands the consequences of the surgery, the lifestyle changes they must adopt afterwards, and that they have realistic expectations about how other aspects of their life may or may not change. In my experience, some people believe that losing weight will automatically improve their relationships and enhance their job performance, and they are often disappointed when this does not happen. I would first recommend psychotherapy. Once the person has developed better skills to regulate their emotions, surgery could then be reconsidered... *Kamila, Clinical Psychology & Psychotherapy*

This point connects with the topics previously mentioned about motivation and expectation and, at the same time, the topic of having enough information to make meaningful choices about one's medical future. Similarly, BID individuals do not seek amputation as a strategy to resolve life problems but to resolve one problem only: the dysphoria caused by the affected limb.

5.5 Conclusions

This chapter explored competency determination within the medical context, emphasising pre-operative risk assessment and post-operative risk behaviour. The pre-operative risk assessment section highlighted the assumption that a lack of decision-making capacity is often inferred when patients ignore medical advice. This behaviour is perceived as contrary to their best interests, revealing a solid paternalistic approach in such cases. Post-operative risk behaviour was defined as the likelihood of patients acting against medical advice, potentially jeopardising the procedure's success. Within this context, impulsivity and emotional stability emerged as crucial factors, distinguishing patients who can make informed medical decisions about their future from those who cannot. The chapter also raised two significant questions: Can a patient's regret after undergoing surgery be anticipated? Moreover, is it more prudent to postpone surgery to understand the condition better and explore potential non-surgical therapies?

Regarding the first point, it was noted that despite the careful assessment of the patient's medical history and psychological assessment, there will never be a certainty that the patient will not engage in unsafe choices that might endanger the post-operative and aftercare phases. Similarly, it was noted that aiming to obtain complete accuracy is a utopia, and as of today, the end goal is to dissipate any reasonable doubt.

For the second point, namely the idea of waiting for future and putative better medical therapies, it was noted how the morality of such a decision might be questionable, assuming the patient satisfies all other criteria (psychological and physical) considered for other invasive medical procedures. In this regard, understanding one's situation, recognising the state of suffering and having the motivation to better oneself to improve one's overall well-being is of paramount importance to be recognised as an autonomous agent.

References

American Psychiatric Association (Ed.). (2013). *Diagnostic and statistical manual of mental disorders: DSM-5* (5th ed.). American Psychiatric Association.

Appelbaum, P. S., & Grisso, T. (1988). Assessing patients' capacities to consent to treatment. *New England Journal of Medicine, 319*(25), 1635–1638. https://doi.org/10.1056/NEJM198812223192504

Bijnen, C. L., Van Den Broek, W. T., Bijnen, A. B., De Ruiter, P., & Gouma, D. J. (2003). Implications of removing a normal appendix. *Digestive Surgery, 20*(3), 215–221. https://doi.org/10.1159/000070388

Buchanan, A., & Brock, D. W. (1986). Deciding for others. *The Milbank Quarterly, 64*(2), 17–94. https://doi.org/10.2307/3349960

Verdugo, J. P. (2022). Rib removal in body contouring surgery and its influence on the waist. *Science & Art in Plastic Surgery Journal, 3*(8144). https://doi.org/10.47511/sapsj.v3.8144

Walker, S. J., West, C. R., & Colmer, M. R. (1995). Acute appendicitis: Does removal of a normal appendix matter, what is the value of diagnostic accuracy and is surgical delay important? *Annals of the Royal College of Surgeons of England, 77*(5), 358–363.

World Health Organization. (2019). *International statistical classification of diseases and related health problems* (11th ed.). Retrieved January 6, 2024, from https://icd.who.int/en

Chapter 6
Moral Responsibility

6.1 Introduction

The research findings indicate that respondents, particularly surgeons, almost unanimously classify on-demand amputations (especially for Body Integrity Dysphoria (BID)) as elective procedures. This classification is significant, as these amputations scarcely meet the conventional criteria for elective surgery. By labelling the procedures as elective, the respondents imply the availability of alternative treatment options for individuals with BID, suggesting that amputation is not considered the only or primary course of action for managing the condition. This perspective introduces the concept of alternate possibilities, thereby raising questions of moral responsibility.

To explore this concept further, this chapter lays the groundwork for understanding moral responsibility by establishing a comprehensive framework. This involves a detailed examination of the value and significance of actions, a clear definition of what qualifies as a moral action and who qualifies as a moral agent, and an in-depth exploration of the nature of moral responsibility itself. With the theoretical framework in place, the subsequent chapter will apply these concepts to BID, examining whether the desire for amputation can be classified as a form of coercion. If it is deemed coercion, the individual would not be considered an autonomous agent. If it is not, two possible avenues to address the reluctance towards amputation emerge: ill will and ignorance.

L. Loriga, *Body Integrity Dysphoria and the Ethical Dilemma of On-Demand Amputation*, https://doi.org/10.1007/978-3-031-82443-2_6

6.2 Assessing the Worth of Actions

One way to interpret every manifested action is by observing the reactions it elicits from others and in ourselves. The response to a particular action hinges on the responder's perception of its value—whether it is positive, meriting praise, negative or attracting blame. An observer evaluates the responsibilities of the actor and responds accordingly, either positively or negatively, based on these assessments. However, such decisions regarding responsibility are subjective and do not offer a comprehensive moral framework for evaluating the action's moral value. The moral value of an action pertains to its ethical significance. This value, or worth, is based on specific criteria: the action's intrinsic nature regardless of its consequences, the consequences themselves, or the character of the acting agent. During discussions about BID with respondents, it became clear that clarifying the moral dimensions of the actions involved was crucial. The primary focus is on two actions: the patient's request for what is believed to be an elective amputation and the medical practitioner's response to this request.

> Moral action is seen, implicitly or explicitly, as complex, imbedded in a variety of feelings, questions, doubts, judgments, and decisions. (Blasi, 1980, p. 1)

Such a perspective frames moral actions as the outcome of intricate cognitive and emotional processes. This array of feelings, questions, doubts, judgements and decisions is inherently anchored in an individual's understanding of particular situations, which in turn shapes their behaviour.

> In any single instance of moral behaviour, understanding guides the action and determines its specific meaning, while action brings moral understanding to its natural completion. (Blasi, 1983, p. 178)

The morality of a given action, therefore, might depend on the sociocultural context within which the action occurs unless one supports the notion of objective morality. In other words, an action deemed moral in one sociocultural context might be considered immoral in another. From this perspective, morality is inherently intertwined with an individual's perception of right and wrong, shaped by personal inclinations, interpretations and sociocultural influences. This suggests that moral judgements are not universally fixed but are instead fluid and context-dependent, reflecting the complex interplay between individual beliefs and societal norms. Such an idea hints towards a specific kind of mental reasoning associated with moral action, what might be called moral cognition (Blasi, 1980, 1983).

> If by moral cognition we mean any verbal statement that has morality as content, three distinguishable kinds of cognitions have traditionally been investigated in their relation to moral action: (a) moral information (i.e., the verbal recognition of moral norms, at least as denned by a specific culture); (b) moral attitudes or values, expressing either a personal belief, an affective inclination, or a tendency to behave in a certain moral manner; and (c) moral judgment or moral reasoning (these two terms are used here interchangeably), characterised by the justification of a moral conclusion and by the general or specific criteria by which moral decisions are supported. (Blasi, 1980, p. 6)

Within this framework, a moral action is defined as any action imbued with moral value. This moral value stems from a process of moral reasoning, which can originate from the individual acting, the sociocultural context in which the action occurs or a combination of both. Thus, the emphasis is not on the intrinsic value of the action itself but instead on its value within a specific sociocultural context. This approach suggests that an action's moral worth is determined by how it aligns with the ethical standards and norms of the surrounding culture and the individual's personal values.

6.3 The Morality of Actions

Not all actions qualify as moral actions, and not every agent qualifies as a moral agent. The attribution of moral worth to an action, whether positive or negative, is pertinent only when the action itself is moral, and the agent performing it is a moral agent. In this context, the concept of moral cognition is particularly useful. Moral cognition, as mentioned already, involves the processes by which individuals perceive, evaluate and decide on moral issues, encompassing the mental activities that underlie ethical reasoning and judgements. It plays a crucial role in identifying who is impacted, prioritising various claims and deciding the best course of action in given circumstances (Blasi, 1980; Greene, 2015; Pratt & Hardy, 2015). These situations can be considered to hold moral significance. A moral situation arises when an individual recognises a moral dilemma that necessitates action based on perceived duty or obligation (Lemmon, 1962). In contrast, non-moral actions do not provoke moral questions or ethical dilemmas. They are viewed as ethically neutral and do not involve considerations of right or wrong. However, the distinction between moral and non-moral actions can often be ambiguous, with certain actions falling into grey areas that are interpreted differently based on cultural, contextual or ethical viewpoints. For this discussion, the classification of an action as moral or non-moral hinges on the individual's subjective perception of a moral dilemma that requires action. This focus on subjective perception is particularly relevant in the context of interviews, where the aim is to understand the respondent's motivations. For instance, a surgeon may face a moral dilemma when deciding whether to perform an amputation requested by a patient, as it involves significant ethical considerations. However, determining the appropriate dosage of a common medication typically does not present the same level of moral conflict.

6.4 Moral Agency and Responsibility

A moral action requires a moral agent as the originating source (with some exceptions McKenna, 2012, pp. 14–20; Zimmerman, 1988, pp. 61–62). A moral agent is an entity that possesses the ability to engage in moral reasoning, make ethical decisions and take actions with moral implications. Such an entity can discern right from wrong and act in alignment with established moral principles and values. For instance, a person who finds a lost wallet might feel a moral obligation to return it to its owner. This sense of duty arises from their capacity for moral reflection and ethical judgement. Conversely, an animal, such as a cat, lacks the cognitive ability to engage in such moral deliberation and is, therefore, not considered a moral agent. As a result, the cat is not subject to moral evaluation or judgement. Any moral agent is responsible for their actions (Fischer & Ravizza, 2008, pp. 1–8) unless their agency is somehow impacted (for some, however, one cannot ever be fully morally responsible Strawson, 1994). Moral responsibility is inherently intertwined with social interactions, manifesting through reactive attitudes such as indignation or admiration towards specific actions. In other words, it is expressed through a range of feelings and attitudes that stem from established interpersonal relationships and are directed towards an entity considered suitable for such responses. This perspective emphasises that moral responsibility is not merely a personal or internal matter but is deeply embedded in the social fabric. It highlights how individuals are judged and responded to within the context of their relationships and the communal standards that govern them. For instance, when someone acts in a way that aligns with or violates shared ethical norms, others respond with feelings of admiration or indignation, respectively. These reactions are not just spontaneous but are rooted in the collective understanding of what constitutes moral behaviour. Thus, moral responsibility involves a complex interplay of personal actions and social expectations, where the moral agent is continually evaluated and held accountable by the community (Strawson, 1962). This paradigm is particularly well-suited to the BID case, as the interviews revealed that respondents, especially surgeons, typically react with scepticism to requests for amputation. Utilising this approach to understand the respondents' reactions enables a comprehensive analysis of how moral responsibility is attributed within this context. The method offers a cost-effective benchmark that aligns with a standardised set of norms, incorporating both general social norms and specific medical standards, thereby ensuring broad and relevant applicability.

6.5 A Particular Type of Reactive Attitude: Blame

Within the framework of this social model of moral responsibility, appraisals can manifest as either blame or praise (or credit). Blame originates from the presumption, whether founded or not, that the agent exhibits ill will when engaging in actions deemed inappropriate by sociocultural standards. Conversely, praise or credit is attributed when actions align with or exceed these sociocultural expectations, reflecting positively on the agent's intentions and ethical considerations.

> To disapprove of him on account of something that he has done or has omitted to do … It is a necessary condition for (morally) blaming someone, in the second sense of 'blame', that

one should believe that he has done something that (morally) one ought not to do or that he has omitted to do something that (morally) one ought to do. (Fields, 1994, pp. 407–408)

The notion of moral disapproval is especially relevant for examining the conflict between elective and non-elective medical treatments, with particular emphasis on the BID case. In these scenarios, ill will may be ascribed, as emerged from the interviews, to either the intention of the acting agent to mislead the physician (who serves as a gatekeeper) to obtain a desired medical intervention or to the patient's lack of understanding about the issue at hand. In this scenario, the attribution of ill will is intensified by the conflict between the physician's perception of the patient's best interests and the patient's request. This presumption of ill will leads to a blaming attitude or moral disapproval. Conversely, praise is regarded as the antithesis of moral disapproval, representing the positive moral evaluation of the agent's actions, thereby recognising them as morally commendable. A moral agent ought to be praised for their actions if their conduct matches or surpasses what any other moral agent would do in the same circumstances.

Consider an individual who witnesses a building fire. If this person rushes to assist those trapped inside, helping as many people as possible escape from the dangerous situation, they would likely be considered praiseworthy among their peers. In contrast, if the same individual ignores the event to avoid being late for a dinner reservation, they would not be seen in the same positive light. Similarly, the individual might still be deemed praiseworthy if they provide aid through other means, such as calling the fire department or alerting nearby residents, as this is an expected response. Within this paradigm, the worthiness of an action is attributed when the moral agent acting has the freedom to choose among various responses (Watson, 1988). The specific choice made by the agent determines the attribution of praise or blame. In this context, *freely* refers to acting autonomously, without internal or external coercion.

The level of autonomy granted by the physician to the patient in the therapeutic relationship is directly influenced by the physician's belief, whether personal or deontological, regarding the elective nature of the patient's request. This insight emerged from the interviews conducted with the surgeon group, highlighting how their perceptions shape the degree of autonomy they are willing to afford patients in these situations. A person seeking medical assistance for what is considered a non-elective condition is generally seen as less blameworthy compared to someone seeking help for an elective condition. For an individual to be deemed deserving of something, they must have considered the likelihood of achieving it before taking any action. This implies that a moral agent must recognise the moral dilemma inherent in a situation and make a conscious decision to act in a particular way, whether good or bad. Conversely, if the individual did not consider the potential outcome due to ignorance, or if they considered it but had no alternative course of action, it is reasonable to conclude that they should not be held fully accountable for the outcome, or their blame should be mitigated.

Consider an individual who causes a significant disruption in a hospital by triggering the fire alarm. Suppose this person was coerced by an armed assailant, who forced them to pull the alarm, leading to panic and potential harm as the hospital staff and patients evacuated. Now, imagine a similar situation where the individual accidentally triggers the alarm because it was poorly marked and mistaken for a door release button or simply because the patient tripped onto some medical equipment left unattended and out of place on the floor and, while falling, grabbed the alarm, triggering it. In the first scenario, it is difficult to blame the individual for the disruption, as the coercion by the armed assailant clearly shifts the fault away from them. In the second scenario, while the individual indeed bears some moral responsibility, the poorly marked alarm (or the unattended medical equipment out of place on the floor) serves as a mitigating factor. It is reasonable to assume that the individual did not foresee causing such a disruption due to the inadequate signage and their unfamiliarity with the hospital layout. Thus, the individual is less blameworthy from a moral standpoint compared to someone who intentionally triggers the alarm with full awareness of the consequences.

Consider now a scenario involving amputation. Suppose an individual amputates their limb, resulting in severe medical conditions that require urgent hospital care. This situation involves significant consequences, such as medical staff prioritising the case over others, doctors working overtime and additional strain on hospital resources. The moral nuances of this act can vary based on the individual's intentions. For instance, if the individual was experiencing a psychotic episode (analogous to being coerced by an armed assailant to trigger the fire alarm from the previous example), it is challenging to attribute full moral responsibility to them. Now, consider a different scenario where the individual inadvertently amputates their limb due to a tragic accident, akin to mistaking a poorly marked alarm for a door release button. Here, while the individual bears some responsibility, the accident serves as a mitigating factor. Lastly, if the individual rationally decides to amputate their limb, believing the outcome would be better than their current situation, the moral responsibility is interpreted differently. In this case, the individual is fully aware and deliberate in their actions, which contrasts with the previous scenarios. Each situation reflects varying degrees of moral responsibility, similar to the different circumstances faced by the fire alarm example. Following this paradigm, it is essential to consider intention, which is the mental state or purpose behind an action (Cushman, 2015; Malle et al., 2014). In the context of BID, evaluating the legitimacy of the request hinges on two key aspects: whether the action was intentional and whether the outcome was intentional (Cohen & Levesque, 1990). When a patient chooses an elective medical procedure, they are supposed to be granted full autonomy, as their decision is made without external pressures, as emerged from the responses gathered throughout the fieldwork phase. In contrast, non-elective medical procedures reduce a patient's autonomy due to internal coercion, stemming from the need to avoid negative consequences if the treatment is not undertaken. This internal coercion affects the patient's ability to choose freely, as their decision is influenced by the necessity to prevent adverse outcomes.

Understanding the difference between elective and non-elective procedures highlights the importance of intention and autonomy in assessing moral responsibility in medical decisions. The patient's intentionality behind their choice and the degree of autonomy they possess are crucial in determining the ethical implications of their decision. Before examining BID, moral responsibility and the roles of intention and coercion, it is essential to establish a foundational understanding of what it means *to do otherwise*. The next section clarifies this foundational concept.

6.6 Choosing an Alternative

> …a person is morally responsible for what he has done only if he could have done otherwise… (Frankfurt, 1969, p. 830)

At the opposite end of the spectrum are scenarios where circumstances not only compel an individual to act but also make avoiding that action impossible. These scenarios may involve external pressures, such as threats of violence, or internal influences, such as false beliefs and misinformation. To fairly evaluate a person's actions, it is essential to understand the specific constraints and limitations they face. Things, of course, are sometimes more complex.

> There may be circumstances that constitute sufficient conditions for a particular action to be performed by someone and that therefore make it impossible for the person to do otherwise, but that does not actually impel the person to act or in any way produce his action. (Frankfurt, 1969, p. 830)

Consider an emergency room doctor who faces a critical shortage of ventilators during a severe health crisis. The doctor is tasked with deciding, based on hospital protocols, how to allocate the limited ventilators between two critically ill patients: one (Patient A) who has a higher chance of survival with the ventilator and another (Patient B) whose condition is more severe and whose outcome is more uncertain. Both patients are in equally urgent need of the ventilator. In this scenario, the shortage of ventilators and the hospital's triage protocols are the necessary conditions that force the doctor to allocate the ventilator to either Patient A or Patient B, but not both. If the shortage were not present, and ventilators were abundant, such a difficult choice would ideally not be necessary. Now, imagine the doctor decides to allocate the ventilator to Patient A, thereby leaving Patient B without this critical support. The causality of this decision stems directly from the lack of resources, compelling the doctor to make a life-and-death choice. However, the decision to prioritise Patient A could also be influenced by the doctor's personal experiences or unconscious biases. For example, the doctor might have lost a close relative to a condition similar to that of Patient A, leading to a stronger inclination to save a life that mirrors their past loss. Alternatively, the doctor might be aware of an incoming supply of alternative treatments that could better serve Patient B's needs than a ventilator—treatments that would be ineffective for Patient A. This knowledge might assure the doctor that both patients could have a chance of survival in the long term, with Patient A receiving the ventilator and Patient B benefiting from the forthcoming resources. Furthermore, the doctor might consider that placing Patient B on a ventilator could provide immediate relief but potentially decrease the patient's resilience to fight for survival when the ventilator is eventually removed. This relaxation of the body's natural defences could, in the long run, be more harmful than beneficial to Patient B. In this context, the doctor could still be considered morally responsible because their actions might have differed if their internal disposition or character had been different, despite the external circumstances leaving them with no practical alternative. Thus, moral responsibility does not always necessitate the ability to choose otherwise.

This chapter does not aim to explore the various interpretations of this dilemma in depth. It is generally accepted, however, that a person coerced into an action is not morally responsible for that action. This understanding supports the paradigm of mutual exclusion between coercion and moral responsibility, which is the framework used here. Freedom and autonomy within this context are framed by the capacity (ought to or can) of a person to do something else (Haji, 1993; Widerker, 1991). In other words, an ability condition must be present. This assessment can be clarified as follows: If someone behaves immorally while acting, they have a moral obligation to refrain from doing it. Failure to refrain from immoral actions renders them blameworthy. If they cannot refrain, then coercion is at play.

6.7 Conclusion

This chapter explored the concept of moral responsibility, particularly in the context of amputation debates. The notable attitude observed among respondents, a natural reaction to emotionally charged and provocative topics, serves as a crucial entry point for understanding the instinctive rejection of on-demand amputation requests. This instinctual reaction, often preceding more considered reflections, provides valuable insights into the moral and ethical reasoning processes of individuals. By shedding light on these inner mechanisms, it becomes clearer why amputation elicits such profound emotional and intellectual turmoil. This understanding can, in turn, clarify the logical gap that separates BID amputation from other similarly invasive medical procedures.

The chapter has also outlined the nature of moral responsibility, emphasising the necessity of a moral framework for assessing the worth of actions. A functional definition of moral actions, moral agents and the complexity of moral cognition has been provided, with an emphasis on the influence of sociocultural contexts. The distinction between moral and non-moral actions has been delineated, highlighting that moral actions are the result of ethical considerations and dilemmas. The concept of moral agency has been defined as the capacity to reason morally and make choices with moral significance. The discussion has also covered the attribution of blame and praise within the social model of moral responsibility, where blame arises from the assumption of ill will. The principle that a person is morally responsible only if they could have chosen an alternative course of action has been established, and the concept of coercion has been examined, emphasising its exclusion from moral responsibility and the importance of assessing internal disposition and character. It is possible now to see how such concepts apply to BID more directly.

References

Blasi, A. (1980). Bridging moral cognition and moral action: A critical review of the literature. *Psychological Bulletin, 88*(1), 1–45. https://doi.org/10.1037/0033-2909.88.1.1

Blasi, A. (1983). Moral cognition and moral action: A theoretical perspective. *Developmental Review, 3*(2), 178–210. https://doi.org/10.1016/0273-2297(83)90029-1

Cohen, P. R., & Levesque, H. J. (1990). Intention is choice with commitment. *Artificial Intelligence, 42*(2), 213–261. https://doi.org/10.1016/0004-3702(90)90055-5

Cushman, F. (2015). Deconstructing intent to reconstruct morality. *Current Opinion in Psychology, 6*, 97–103. https://doi.org/10.1016/j.copsyc.2015.06.003

Fields, L. (1994). Moral beliefs and blameworthiness: Introduction. *Philosophy, 69*(270), 397–415. https://doi.org/10.1017/S0031819100047239

Fischer, J. M., & Ravizza, M. (2008). *Responsibility and control: A theory of moral responsibility*. Cambridge University Press.

Frankfurt, H. G. (1969). Alternate possibilities and moral responsibility. *The Journal of Philosophy, 66*(23), 829–839. https://doi.org/10.2307/2023833

Greene, J. D. (2015). The rise of moral cognition. *Cognition, 135*, 39–42. https://doi.org/10.1016/j.cognition.2014.11.018

Haji, I. (1993). Alternative possibilities, moral obligation, and moral responsibility. *Philosophical Papers, 22*(1), 41–50. https://doi.org/10.1080/05568649309506392

Lemmon, E. J. (1962). Moral Dilemmas. *The Philosophical Review, 71*(2), 139–158. https://doi.org/10.2307/2182983

Malle, B. F., Guglielmo, S., & Monroe, A. E. (2014). A theory of blame. *Psychological Inquiry, 25*(2), 147–186. https://doi.org/10.1080/1047840X.2014.877340

McKenna, M. (2012). Moral responsibility for conduct. In *Conversation and responsibility* (pp. 14–21). Oxford University Press. https://doi.org/10.1093/acprof:oso/9780199740031.001.0001

Pratt, M., & Hardy, M. (2015). Cultivating the moral personality: Socialization in the family and beyond. In J. E. Grusec & P. D. Hastings (Eds.), *Handbook of socialization: Theory and research* (2nd ed., pp. 661–685). The Guilford Press.

Strawson, P. (1962). Freedom and resentment. *Proceedings of the British Academy, 48*, 187–211.

Strawson, G. (1994). The impossibility of moral responsibility. *Philosophical Studies: An International Journal for Philosophy in the Analytic Tradition, 75*(1/2), 5–24.

Watson, G. (1988). Responsibility and the limits of evil: Variations on a Strawsonian theme. In F. Schoeman (Ed.), *Responsibility, character, and the emotions: New essays in moral psychology* (pp. 256–286). Cambridge University Press.

Widerker, D. (1991). Frankfurt on 'ought implies can' and alternative possibilities. *Analysis, 51*(4), 222–224. https://doi.org/10.1093/analys/51.4.222

Zimmerman, M. J. (1988). Appraisability. In *An essay on moral responsibility* (pp. 38–73). Rowman & Littlefield.

Chapter 7
Blame and Amputation: Uncovering Preconceptions and Addressing Physicians' Scepticism

7.1 Introduction

This chapter continues to analyse physicians' attitudes, as revealed through the interviews and fieldwork, with a focus on a new aspect: the worthiness of the chosen action path to inquire into the ideas of voluntary or involuntarily deceitfulness of the patient to obtain types of interventions that extend beyond current medical standards. One central question drives this discussion: Is seeking amputation through official medical channels a moral action, and should it be considered praiseworthy or blameworthy? This analysis is crucial for understanding the scepticism physicians exhibit towards on-demand amputation and similar issues. As the fieldwork and interviews revealed, their reluctance is rooted in the belief that individuals with Body Integrity Dysphoria (BID) who seek amputation are perceived as more blameworthy than those undergoing non-elective procedures driven by medical necessity. The concept of blameworthiness is rooted in the assumption of something superfluous and dangerous. Such an assumption raises questions regarding the motivations behind a patient's actions and hints that a covert ill will may be driving their behaviour. Rather than seeking relief from suffering, the patient may desire personal satisfaction, whether from voluntary inclinations or involuntary impulses, as might be the case in paraphilic disorders. Such potential attribution of blame hinges on the premise that medical knowledge and expertise are being used as tools to achieve improper ends. This interpretation implies that the patient uses medical means not to improve or restore their health but for purposes that deviate from an accepted medical and social framework.

It is essential to recognise that this interpretation of blameworthiness is not the sole lens through which such cases were understood by the respondents. Another is that the patient may lack awareness of other possible options. Such an approach shifts the focus from intentional ill will towards a lack of knowledge. Patients might make blameworthy choices due to their unawareness of more appropriate courses of

L. Loriga, *Body Integrity Dysphoria and the Ethical Dilemma of On-Demand Amputation*, https://doi.org/10.1007/978-3-031-82443-2_7

action. This lack of knowledge may be attributed to a patient's limited access to information, societal stigmas surrounding certain conditions, or even a scarcity of medical resources to explore other solutions. Recognition of the potential duality of blameworthiness—intentions driven by improper ends or a lack of knowledge—enhances the understanding of the factors influencing medical decision-making.

Finally, the chapter presents thought-provoking reasoning on the moral implications of self-made amputations. It explores the extent to which pursuing amputation through unofficial channels renders the individual morally responsible for their actions and why this is a significant consideration. The chapter then concludes with the contentious argument that, in some cases, a self-made solution might actually represent a praiseworthy moral choice for individuals with BID.

7.2 Application of Moral Responsibility to BID

Two core points must be considered to gain a better understanding of BID from a moral perspective: coercion and intention. These points are crucial in framing a BID person's desire for impairment. Addressing these aspects is essential to clarify the moral status of the phenomenon and to evaluate whether the scepticism of medical personnel regarding the requested amputation is justified.

7.2.1 Is the Desire of Amputation Coercive?

Coercion refers to the limitation or removal of an individual's capacity to choose freely. It involves external pressures that compel a person to engage in actions or behaviours they would otherwise avoid under normal circumstances. In the context of BID, the intense desire for impairment acts as a coercive force, driving individuals to pursue actions, such as amputation, that they might not consider if not for this overpowering influence. The interpretation, understanding and reporting of such desires have evolved significantly over the years. From an urge which can "...suddenly arrive full force upon the individual who can then be overwhelmed with desire ... it is so deeply ingrained within the apotemnophilia that she/he cannot restrain or regulate its direction" (Furth & Smith, 2002, p. 68) to more moderate definitions that frame these desires as strong but manageable impulses akin to "the intense and persistent desire" of the ICD-11 (World Health Organization, 2019, 6C21). Clarifying once and for all the exact intensity of such desire in this context would be presumptuous. However, it is fair to say that, in certain instances, the desire is strong enough to drive individuals to attempt self-mutilation (Bensler & Paauw, 2003; Berger et al., 2005; Blancard, 1991; Brenner, 1999; Dyer, 2000; First, 2005; Sorene et al., 2006; Stewart & Lowrey, 1980; Tavcar et al., 1999).

Depending on the interpretation, this desire may cloud an individual's rational judgement, compelling them towards a specific action even if they would have preferred to choose differently. If this desire is accepted as coercion, the individual with BID is seen as lacking moral responsibility for the consequences. The intense desire for impairment is considered a coercive force, thus stripping the individual of autonomous decision-making and disqualifying the request for amputation outright. Conversely, if the impairment desire is not viewed as coercion, the individual with BID is regarded as an autonomous agent capable of making alternative choices. In this case, they have the capacity to make informed decisions about their medical future, which introduces the possibility of moral blameworthiness. A third interpretation is also possible. Here, the wish for impairment fluctuates in intensity, sometimes overwhelming the individual and placing them in a state of coercion and, at other times, diminishing, allowing for autonomous decision-making. This variability can lead to intermittent periods where the individual oscillates between coerced and autonomous states.

The first crucial step in addressing this issue is to establish precise definitions. Understanding whether the desire associated with BID qualifies as an obsession requires a clear definition of obsession itself. Despite the availability of multiple definitions, they all generally converge on two fundamental points: an obsession is characterised by an uncontrollable compulsion, and it substantially diminishes an individual's autonomy.

> Obsession can be defined as a burdensome and uncontrollable compulsion to repeat an idea, emotion, or thought ... accompanied by feelings of distress or anxiety in the consciousness of the obsessive person … While obsession is often considered pathological, such a definition is possible only if the obsession obstructs the free will of the person in a prohibitive and disquieting way. (Hourigan, 2011, pp. 1056–1057)

Similarly, obsessions are "recurrent and persistent thoughts, impulses or images that are experienced as intrusive and inappropriate and that cause marked anxiety or distress" (Oulis et al., 2013, p. 51) and "repetitive and persistent thoughts (e.g., of contamination), images (e.g., of violent scenes), or impulses/urges (e.g., to stab someone) that are experienced as intrusive, unwanted and are commonly associated with anxiety" (World Health Organization, 2019, MB26.5). Gaining a clear understanding of compulsion is also essential for interpreting the concept of obsession within the context of BID.

> Repetitive behaviours or rituals (e.g., washing, checking) or mental acts (e.g., repeating words silently) that the individual feels driven to perform in response to an obsession, according to rigid rules, or to achieve a sense of 'completeness.' (World Health Organization, 2019, MB23.4)

These definitions bring the concept of BID closer to that of obsessive-compulsive disorder (World Health Organization, 2019, 6B20). However, compulsion is not explicitly mentioned in the definition of BID (World Health Organization, 2019, 6C21), despite the notable similarity in the language used to describe both conditions. After confirming that the desire in BID does not correspond to a compulsive disorder (as per the official definition of BID), the next objective is to identify the origins of this desire and its effects on the individual.

While the exact cause of BID is currently unknown, it is helpful to differentiate its underlying factors from those associated with obsessive disorders. Both BID and obsessive disorders can result from a mix of genetic, environmental and developmental influences. However, a significant distinction is that obsessive disorders often involve irrational and contradictory obsessions (Bozorgmehr et al., 2017), whereas BID consistently involves a singular and pervasive sense of bodily overcompleteness. At this point, one might question whether the individual with BID desires amputation itself or seeks to eliminate their suffering, which they believe amputation will resolve. It is reasonable to propose that what appears to be an obsession in the context of BID actually stems from a long-standing, rational desire.

Consider a person suffering from chronic, debilitating migraines. This individual might develop an intense preoccupation with finding relief, which could appear obsessive. Although this preoccupation exhibits characteristics of an obsession—persistent and repetitive thoughts about alleviating pain—it would be unreasonable to deem such a focus irrational. Applying this reasoning to BID suggests that prolonged suffering might lead to a similar fixation on amputation. Labelling the desire for amputation as irrational overlooks the fact that, in this context, rationality involves seeking relief from suffering rather than enduring it. The apparent rationality of the desire for amputation in BID might become obscured when the desire reaches its peak intensity. If likened to a wave, the individual might be driven to extreme actions when the wave is at its highest point. Conversely, rationality is more likely preserved when the wave is at its lowest. Like a wave, the cycle of the BID desire from crest to trough affects its intensity and, consequently, the rationality of the desire.

Another point to consider is the drive individuals have to improve a current state of dissatisfaction, striving for self-betterment. This drive can be seen as a rational decision to change their circumstances, akin to the rationale observed in obese patients seeking gastric banding, as noted in the previous sections. Based on this understanding, the desire stemming from BID should not be viewed necessarily as an obsession that undermines personal autonomy. However, this desire may exert a coercive influence, pushing the individual towards seeking relief, through unorthodox ways. Ignoring this drive might be seen as less rational than acting on it, especially when the ultimate goal is to enhance one's well-being. Thus, the desire experienced by someone with BID, does not necessarily compromise their moral responsibility, especially when the intensity of the desire is low or increasing. When such a desire is acknowledged as legitimate under current medical standards, the individual should not be judged harshly. Instead, they should be viewed as acting rationally, much like anyone seeking to escape discomfort. Persisting in a continuous state of suffering and dissatisfaction, without seeking a solution, could more reasonably be seen as irrational.

7.2.2 Does the Intention Matter?

Having established that certain conditions can prevent recurrent and persistent thoughts from being classified as obsessions and recognising that some obsessions may have rational foundations that do not negate an individual's moral responsibility, the focus can now shift to evaluating the legitimacy of the expressed behaviours or actions. The next step is to decipher the possible intentions behind the actions of individuals with BID. Understanding these intentions is essential for evaluating the ethical worthiness of actions associated with BID. By delving into the motivations of individuals with BID, it becomes possible to assess the moral implications of their desire for amputation more accurately. This analysis allows for the determination of whether these actions are rational responses to their condition or if they require a different kind of ethical consideration. The pursuit of amputation within the BID context, as revealed by the interview pool, is often believed to originate from two distinct factors—ill will and ignorance—each carrying drastically different implications depending on which is taken into account. Two central questions emerge: Is the goal to achieve amputation itself, or is it to relieve suffering by any suitable means? These scenarios may not necessarily overlap. The following section will answer the question of what is meant by suitable means. Now, however, it is necessary to understand that the pursuit of amputation might stem from different underlying motivations, each carrying significantly different consequences.

The first factor to consider is whether the amputation request stems from the individual's conscious intention to deceive the physician in order to achieve a goal different from the one initially presented during the consultation phase. In other words, the narrative surrounding the need for amputation could be a façade used to secure an amputation for reasons not disclosed to the physician. This paradigm suggests deceitfulness. The motivations for such deceit could be legitimate or illegitimate. For instance, consider a hypothetical condition more severe than BID, not yet recognised by any official medical taxonomy, but with a similar pathogenesis. An individual suffering from this condition might see amputation as a significant relief, similar to how it might benefit someone with BID. In this scenario, if the amputation alleviates immediate suffering and prevents potentially life-threatening complications without causing more harm than a BID-related amputation would, the individual's actions are justified from a personal perspective. However, the act of lying and deceiving the physician remains ethically problematic, even if the underlying reasons seem legitimate. The same can be speculated for less justifiable reasons, such as paraphilic disorders, where atypical sexual arousal is linked to specific bodily configurations. In essence, when a physician is unable to evaluate the true reasons behind an amputation request due to the individual's deceitfulness, the request itself is tainted. Despite any potentially legitimate motivations, the individual can be deemed blameworthy. Applying this to the initial statement, the individual, considering the likelihood of obtaining an amputation, chose to follow a course of action involving dubious practices.

Assessing the worthiness of the individual and their action may initially seem straightforward, but it requires clarification to consider the second factor properly. If there is no malicious intent but rather ignorance of their condition and available therapeutic approaches, is the individual seeking amputation still deemed blameworthy? In other words, can ignorance serve as an excuse for moral responsibility? To address this, it is essential to explore whether the individual's lack of knowledge about their condition and possible treatments absolves them of moral responsibility. Suppose the individual genuinely does not understand the full scope of their condition or the alternative treatments available. In that case, their actions may not be driven by deceit but by a misguided attempt to find relief. Ignorance, in this context, could potentially mitigate blameworthiness. If an individual sincerely believes that amputation is the only solution to their suffering due to a lack of information, their actions may be viewed more sympathetically. The absence of ill intent shifts the focus away from moral culpability. However, ignorance alone does not entirely absolve moral responsibility. There is still an expectation that individuals seek adequate medical advice and explore all available options. The healthcare system also bears a responsibility to ensure that patients are fully informed about their conditions and the potential consequences of their choices. Therefore, while ignorance might reduce the degree of blame, it does not completely eliminate moral responsibility. The individual's actions must be evaluated within the context of their knowledge and the efforts made to understand and address their condition. A possible approach to the problem of ignorance is the following:

> Ignorance of the universal is ignorance of what one ought to aim at and what one ought to do or abstain from. Ignorance of this kind may be described as moral ignorance. Ignorance of particulars is ignorance of the circumstances of the action and the objects with which it is concerned. (Fields, 1994, p. 397)

Now, an individual is relieved of moral responsibility only in cases of ignorance of particulars, where the individual is not responsible for their lack of knowledge. This holds as long as the individual genuinely had no opportunity or ability to acquire the necessary knowledge. Suppose an individual had the chance and means to understand their condition and available treatments but chose not to pursue this knowledge. In that case, they can be held accountable for their ignorance. In other words, moral responsibility is mitigated only if the individual's ignorance is not self-inflicted. When an individual has access to information and the capacity to seek it out, yet fails to do so, they remain culpable for their ignorance. The critical distinction lies in whether the ignorance is involuntary or a result of negligence. On a similar note, an agent is morally ignorant if one

> fails to know or truly believe that actions of a certain sort are wrong or that they ought not to be done. That an agent fails to know or truly believe that actions of a certain sort are wrong does not necessarily imply, but is nevertheless compatible with, his believing that actions of that sort are right. (Fields, 1994, p. 398)

It also follows that:

> If a person does what he thinks is right, and his thinking so depends on his belief that actions of that sort are right (whereas, in truth, actions of that sort are wrong) he does not …

> deserve blame. For whether or not a person deserves blame depends not on what he does but on what he intends to do. In the case in question, the agent acts intending to do what is right. (Fields, 1994, p. 398)

This implies that actions can be wrong while intentions remain good (due to the agent's lack of knowledge), or intentions can be wrong while actions might be right. In the context of BID, this distinction is relevant when considering ill will versus good intentions. Ill will relates to intention, while deceptiveness pertains to blame. If ill will is absent, the individual's intentions are correct, even if the actions are perceived as wrong. In such cases, the individual may not necessarily be held morally responsible or blameworthy. A person with BID might be ignorant of the specific circumstances and the nature of the actions they seek. This ignorance, particularly regarding the implications of amputation, cannot easily be attributed to the individual, given the unclear aetiology of BID and the contentious nature of its treatment. Thus, while the actions of a person with BID might be deemed misguided or inappropriate, their intentions could still be considered well-meaning. The lack of clear medical understanding and therapeutic consensus around BID further complicates the issue, making it unfair to hold the individual fully accountable for their ignorance. Nonetheless, using an individual's lack of knowledge of specific circumstances to absolve them of moral responsibility is a weak argument unless it is evident that their ignorance is due to factors beyond their control. If uncontrollable forces are at play, suggesting potential coercion, these forces could prevent the individual from acquiring the necessary knowledge, thus relieving them of moral responsibility. However, excluding such coercive forces for the sake of argument, ignorance might mitigate an agent's responsibility but not entirely absolve it. The impact of this lack of knowledge on the individual's decision-making remains ambiguous. Certain aspects of their situation, such as understanding the consequences of their actions, their responsibilities to themselves and others and legal requirements, are fundamental knowledge that individuals are expected to possess.

In the context of BID, while ignorance regarding the full implications of amputation might lessen the blame, it does not eliminate responsibility entirely. Individuals are generally expected to seek out information that impacts their well-being and decision-making processes. Therefore, while a person with BID may not be wholly responsible for their actions due to their lack of knowledge, they are not entirely free from moral responsibility either. The degree to which this ignorance affects their moral culpability is complex and context-dependent. It requires a careful examination of the factors that contributed to their lack of knowledge and the extent to which they could have reasonably been expected to overcome these barriers. People with BID who seek amputation are not entirely morally responsible for their requests due to their ignorance of the specific context. If their moral intentions are sound (i.e. they are not ignorant of the universal), they might still lack the necessary understanding to make an informed decision. Such understanding includes comprehending the extensive implications of undergoing a major surgery like limb removal. This encompasses awareness of the financial burden of post-operative therapies not covered by statutory health insurance, for example, potential social stigma, and the impact on social relationships. While this perspective appears reasonable, it raises another important question: Why is seeking amputation considered the wrong action in this context? Conversely, under what circumstances might amputation be considered the right action?

7.2.3 Why Not Amputation?

If ill will is absent, the actions of individuals with BID seeking amputation are not entirely blameworthy, as long as they do not stem from moral ignorance. The argument shifts if multiple medically recognised therapeutic options for BID are available, excluding amputation, and the patient persists in requesting amputation against professional medical advice. The existence of such alternatives could invoke Fields' exception regarding ignorance of particulars. Blame is lifted only if the individual is not responsible for their ignorance.

Suppose individuals have access to information about alternative treatments and the potential consequences of amputation, but they choose to ignore or reject this knowledge. In that case, they bear some responsibility for their decision. The insistence on an unsupported medical option, despite the existence of legitimate alternatives, reintroduces a degree of blameworthiness. Therefore, while the absence of ill intent reduces blame, the availability of recognised treatments for BID complicates the ethical considerations.

One potential issue with this formulation is the disparity in information access between the patient and the medical professional. It is neither reasonable nor accurate to expect the patient, rather than the physician, to suggest a therapeutic approach. Typically, individuals affected by a disease or illness are not equipped to diagnose their condition and determine the appropriate solution independently. Such a scenario would be not only unrealistic but would also render the medical profession somewhat obsolete. The critical question, then, is how to assess the patient's understanding of what it means to live post-operatively with such a physical impairment. Equally, it is plausible to consider that the request for amputation arises from the patient's identification of the source of their discomfort somewhere in their body, leading them to instinctively want to remove that source. When the discomfort is localised to a limb, the request for amputation—potentially stemming from the patient's ignorance—appears more understandable and logical than previously assumed.

This perspective allows for addressing the second part of the initial problem: the possibility of doing otherwise. The critical question now is whether individuals with BID can, at present, pursue alternatives to amputation and what these alternatives entail in this context. Understanding the motivations behind a patient's request for amputation involves recognising the instinctive desire to eliminate the perceived source of suffering. While perhaps influenced by a lack of comprehensive knowledge, this request reflects a logical response to the discomfort felt in a specific part of the body. The challenge then becomes examining what *doing otherwise* means for individuals with BID and whether viable alternatives exist. *Doing otherwise* encompasses exploring different therapeutic approaches that do not involve amputation. These alternatives could include psychological interventions, pharmacological ones or other medical treatments aimed at addressing the underlying causes of BID. It is reasonable to assume that any individual experiencing significant suffering or illness would seek relief by any means necessary. To do otherwise—that is otherwise than amputate—would imply one of three possibilities: the suffering is not severe enough to cause substantial discomfort, the individual is compelled to remain in such a state due to coercion or restrictive forces, or the individual is acting irrationally.

People with BID are affected by suffering or illness. As noted, they are considered rational despite their unusual desire. It is reasonable to assume that an individual with BID would seek relief by any means necessary. Currently, there are three means at the disposal of that individual: continuing to suffer by preserving the status quo, seeking medical aid and finding self-made solutions. Enduring the suffering caused by BID diminishes the quality of life and places individuals outside the spectrum of autonomy identified in the earlier debate. This scenario leaves two potential paths: medical intervention and self-devised solutions. Considering that no therapeutic approach has proven effective for treating BID other than amputation (Greene, 2015; Kröger et al., 2014), surgical removal of the limb may be viewed as a legitimate means to achieve the desired outcome.

Consider a scenario where alternate therapies are available that achieve the same results as amputation without involving any irreversible surgical procedures. Under these circumstances, amputations can be considered elective. This reasoning extends to the treating physician. Performing an amputation is blameworthy for the same reasons an individual with BID seeking amputation might be blamed. Conversely, pursuing alternative treatments is praiseworthy for several reasons, including the unclear aetiology of BID and the lack of established therapeutic approaches. Thus, the blameworthiness in the BID case stems from the individual's choice to seek amputation when other options are available. When these alternatives do not exist, and medical aid is unavailable, individuals are left with self-devised solutions. This reasoning can be further extended for the sake of argument. Given the lack of available therapies, remaining in a state of suffering can be viewed as a form of coercion in the absence of medical aid. From this perspective, self-made solutions become a viable alternative.

> If a person intends to do what is right and acts accordingly and, in doing so, has to overcome a strong temptation to do otherwise, then not only is he not blameworthy; he is praiseworthy. For he has done what he conceives of as his duty despite a strong personal inclination to act differently… (Fields, 1994, p. 398)

Applying this reasoning to the BID case leads to the conclusion that individuals with the condition, compelled to escape their suffering and overcome the fundamental instinct of self-preservation by mutilating their bodies for significant hypothetical benefits, deserve praise for their actions. This perspective supports the advancement of self-made solutions and highlights a further dilemma concerning the BID phenomenon.

7.3 Conclusion

This chapter concluded the debate on moral responsibility initiated earlier. Moral responsibility encompasses the capacity to understand and reason about moral issues (moral cognition) and the willingness to act in ways deemed morally right or just in morally significant situations. Within this framework, blame was examined as an epistemic state that emerged from the empirical data gathered through extensive interview pools when debating amputation for BID individuals with medical professionals. Two key questions were addressed: whether individuals with BID qualify as moral agents despite their predominant desire for impairment and whether their pursuit of amputation is an action worthy of blame. Establishing a foundational understanding of moral responsibility and the criteria for excluding an agent from it, coercion and intention were highlighted as critical factors. Desire arising from BID may or may not constitute coercion, depending on the interpretative approach. However, this desire, stemming from a rational need to alleviate suffering, does not inherently disqualify individuals from possessing rationality and moral responsibility.

Intention was analysed in two contexts: the intention to achieve amputation and the intention to relieve suffering. Clinicians involved in this research distinguish between patients with ill intentions, who deceive physicians to achieve a specific goal, and those unaware of the full implications of their request. In the first scenario, the patient may be deemed blameworthy; the second remains open to debate. Referencing Fields' distinction between ignorance of the universal and ignorance of particulars, the BID phenomenon was examined to identify similarities and discrepancies with the gathered data. Excluding ill will leaves ignorance of particulars, which somewhat absolves individuals with BID of blame. Consequently, seeking medical amputation appears not blameworthy but a rational step given the circumstances.

Further speculation suggests that, in the absence of available medical therapies, self-administered amputations can be understood and even commended. This argument aims to offer a provocative perspective on the complex issues surrounding BID. While self-administered amputation is not ideal, individuals with BID, like any rational agents, strive to improve their condition by any means necessary, even if it means overriding their instinct for self-preservation. This dynamic places individuals with BID in a difficult position, emphasising the urgent need for less destructive treatment options. Taking no definitive stance in this context cannot be considered the least harmful solution.

References

Bensler, J. M., & Paauw, D. S. (2003). Apotemnophilia masquerading as medical morbidity. *Southern Medical Journal, 96*(7), 674–676. https://doi.org/10.1097/01.SMJ.0000078367.94479.B9

Berger, B. D., Lehrmann, J. A., Larson, G., Alverno, L., & Tsao, C. I. (2005). Nonpsychotic, nonparaphilic self-amputation and the internet. *Comprehensive Psychiatry, 46*(5), 380–383. https://doi.org/10.1016/j.comppsych.2004.12.003

Blancard, R. (1991). Clinical observations and systematic studies of autogynephilia. *Journal of Sex & Marital Therapy, 17*(4), 235–251. https://doi.org/10.1080/00926239108404348

Bozorgmehr, A., Ghadirivasfi, M., & Shahsavand Ananloo, E. (2017). Obsessive–compulsive disorder, which genes? Which functions? Which pathways? An integrated holistic view regarding OCD and its complex genetic etiology. *Journal of Neurogenetics, 31*(3), 153–160. https://doi.org/10.1080/01677063.2017.1336236

Brenner, I. (1999). Upper-extremity self-amputation in a case of dissociative identity disorder. *The Journal of Clinical Psychiatry, 60*(10), 705–706. https://doi.org/10.4088/JCP.v60n1011

Dyer, C. (2000). Surgeon amputated healthy legs. *BMJ, 320*(7231), 332. https://doi.org/10.1136/bmj.320.7231.332

Fields, L. (1994). Moral beliefs and blameworthiness: Introduction. *Philosophy, 69*(270), 397–415. https://doi.org/10.1017/S0031819100047239

First, M. B. (2005). Desire for amputation of a limb: Paraphilia, psychosis, or a new type of identity disorder. *Psychological Medicine, 35*(6), 919–928. https://doi.org/10.1017/S0033291704003320

Furth, G., & Smith, R. (2002). *Amputee identity disorder: Information, questions, answers and recommendations about self-demand amputation*. Authorhouse.

Greene, J. D. (2015). The rise of moral cognition. *Cognition, 135*, 39–42. https://doi.org/10.1016/j.cognition.2014.11.018

Hourigan, D. (2011). Obsession. In D. Southerton (Ed.), *Encyclopedia of consumer culture*. SAGE Publications, Inc. https://doi.org/10.4135/9781412994248

Kröger, K., Schnell, T., & Kasten, E. (2014). Effects of psychotherapy on patients suffering from body integrity identity disorder (BIID). *American Journal of Applied Psychology, 3*(5), 110–115. https://doi.org/10.11648/j.ajap.20140305.11

Oulis, P., Konstantakopoulos, G., Lykouras, L., & Michalopoulou, P. G. (2013). Differential diagnosis of obsessive-compulsive symptoms from delusions in schizophrenia: A phenomenological approach. *World Journal of Psychiatry, 3*(3), 50–56. https://doi.org/10.5498/wjp.v3.i3.50

Sorene, E. D., Heras-Palou, C., & Burke, F. D. (2006). Self-amputation of a healthy hand: A case of body integrity identity disorder. *Journal of Hand Surgery, 31*(6), 593–595. https://doi.org/10.1016/J.JHSB.2006.05.022

Stewart, D. E., & Lowrey, M. R. (1980). Replantation surgery following self-inflicted amputation. *The Canadian Journal of Psychiatry, 25*(2), 143–150. https://doi.org/10.1177/070674378002500208

Tavcar, R., Dernovsek, M. Z., & Zvan, V. (1999). Self-amputation of left hand: A case report. *The Journal of Clinical Psychiatry, 60*(11), 793–794. https://doi.org/10.4088/JCP.v60n1117

World Health Organization. (2019). *International statistical classification of diseases and related health problems* (11th ed.). Retrieved January 6, 2024, from https://icd.who.int/en

Chapter 8
Health and Well-Being: Philosophical Approaches

8.1 Introduction

Seeking on-demand amputation does not disqualify the individual from having moral responsibility. Similarly, amputation is not inherently a blameworthy practice. Instead, it can be used provocatively to advocate the rationality of the Body Integrity Dysphoria (BID) individual because amputation is, as of today, the only way available for such individuals to avoid a perpetual state of suffering. Such an initiative from the patient's side, can be used to infer the rationality, self-reflection and medical decision-making capacity. Such formulation also leverages the assumption that to seek relief, the individual must go against one's self-preservation instinct, which is praiseworthy as it defies accusations of impulsivity. In other words, seeking amputation aims to achieve a hypothetical state of well-being, which the BID individual lacks in the current condition.

To fully comprehend the rationale behind the desire for amputation in individuals with BID and to provide a foundation for supporting the BID perspective (though this is not the primary aim), it is essential to engage deeply with the concept of well-being. Introducing well-being into the discussion of BID is critical, as it is predicated on the fundamental idea that specific goals typically drive individual behaviours. It is difficult to argue that people navigate life without some sense of purpose, direction or objectives. These goals are inherently shaped by a person's unique inclinations, life experiences and the cultural and geographical contexts in which they live. For individuals with BID, the desire for amputation can be understood as a health-related goal, one that is deeply intertwined with their conception of well-being. A goal can be defined as *"a desirable future state of affairs one intends to attain through action"* (Kruglanski, 1996, p. 600).

This chapter seeks to develop a comprehensive model of well-being capable of justifying the request for amputation in individuals with BID. In essence, this model aims to determine whether amputation serves the best interests of BID patients from

L. Loriga, *Body Integrity Dysphoria and the Ethical Dilemma of On-Demand Amputation*, https://doi.org/10.1007/978-3-031-82443-2_8

a well-being perspective. The chapter begins by addressing key terminological issues to ensure clarity in subsequent discussions. It then proceeds to explore three prominent theories of well-being, offering a detailed examination of how well-being is currently conceptualised and debated within the field of bioethics. The analysis will demonstrate that any of the three theoretical frameworks can support the decision to accept or reject amputation, contingent upon the interpretative approach applied to each paradigm.

8.2 What Is Well-Being

8.2.1 Terminology

The idea of well-being is broad. Well-being *"encompasses all the ways in which people experience and evaluate their lives positively"* (Tov, 2018, p. 1), which highlights the subjective and interpretative nature of the concept. Well-being can also represent a state in which what is good for the individuals is rooted in fulfilling their desires. Conversely, what is bad for individuals is derived from their inability to achieve what they want or hindering their desires (Heathwood, 2017, pp. 135–137). Alternatively, well-being can be none of the above; instead, it can represent the state obtained by promoting specific objective elements such as friendship, wealth and love, regardless of whether the individual recognises them as intrinsic good or bad. Before progressing further in the discussion however, it is essential to address another foundational concept: health. During the fieldwork phase, a practical definition of health emerged, which can serve as a useful guideline. Health is conceptualised as the optimal functioning of the organism, where every component works in harmony to contribute to the individual's overall functionality. This perspective defines health as a state marked by the absence of dysfunction, where the various parts of the organism synergistically support the individual's holistic functioning. In contrast, a decline in health is observed when there is a deviation from an established hypothetical norm, leading to diminished functionality. The complexity of these concepts creates confusion regarding how well-being should be defined and assessed. Further elaboration is required to make these concepts more usable for debate. The first step in this process is to clearly differentiate between objectivist and subjectivist approaches to well-being by providing precise definitions, thereby laying the groundwork for a more nuanced discussion of BID.

8.2.2 Objective Versus Subjective Stances

For this discussion, subjective theories of well-being will be defined as those in which a person's welfare is primarily assessed based on their mental state, with the individual's subjective experience serving as the key determinant of their well-being. These theories emphasise that well-being is closely tied to how individuals perceive and feel about their lives, making personal attitudes, preferences and mental states central to evaluating well-being. In contrast, objective theories of well-being will be characterised as those in which an individual's well-being is primarily determined by factors external to their mental state. From this perspective, well-being is seen as something that can be evaluated and measured against external criteria, independent of the individual's personal attitudes, preferences or subjective experiences. These theories assert that certain objective conditions or goods are necessary for well-being, regardless of how the individual perceives them. The subjectivist paradigm, therefore, opposes this view, advocating for an attitude-dependent approach where the individual's mental state is the primary criterion for assessing well-being. This distinction between subjective and objective approaches is crucial for understanding the differing perspectives on well-being and how they can be applied in cases such as BID.

8.2.3 What Is Well-Being

Well-being can encompass various elements, including physical health, mental and emotional health, strong and fulfilling social relationships and financial stability. Although it is hard to argue that any of these elements do not represent well-being, it is similarly challenging to sustain that the idea of well-being falls exclusively within one of these domains. Instead, well-being is a multifactored ideology branching across several grounds. Overall, chief theories of well-being fall into three categories (see, for example, Parfit, 1986, p. 493). These three approaches are Hedonism, Desire-fulfilment theories and Objective-list theories. The following sections will provide an examination of these theories. The analysis will delve into their fundamental principles, key concepts and underlying assumptions, setting the stage for a debate on BID.

8.3 Hedonism

Hedonism is the stance par excellence for evaluating an individual's life from a subjective perspective. Such an approach positions the individual at the centre of the assessment, focusing on self-evaluation as a critical element in understanding one's sense of well-being. Within such a paradigm, the individual is called to judge oneself strictly from a personal perspective. Such judgment entails both an affective level and a cognitive one. Similarly, such a judgment is measured against strictly personal criteria. Such criteria are framed against positive emotions, the absence of negative emotions and the intricate interplay between these emotional aspects to reach maximum pleasure.

8.3.1 Arguments for Hedonism

The advantages of adopting Hedonism in assessing well-being are evident, the first being that the individual has knowledge of oneself, which any external observer or normative criteria can hardly match. One individual might enjoy eating sugary treats, while another might despise their taste. Assuming sugary treats are good for everybody would be a fallacious assumption. Similarly, one can appreciate having a specific body type—being tall, short or having an uneven number of limbs as for BID—while another cannot even comprehend how such a specific bodily configuration can be considered fulfilling. Here, Hedonism offers an advantage by tailoring well-being to specific individuals. Instead of sweeping generalisations about what brings well-being to all, Hedonism acknowledges and embraces the diversity of human experiences and preferences, ensuring that well-being is individualised and aligns with the pleasure or satisfaction each person seeks. Adopting Hedonism to enhance one's understanding of BID is undeniably appealing, given its alignment with the distinctive nature of the BID experience in terms of health-related goals.

However, it is important to highlight that solely depending on one's subjective perspective as the sole metric for evaluating well-being can give rise to self-referencing issues. The dilemma here is that individuals grappling with BID might find themselves entangled in a self-contained feedback loop in which the emotional experiences concerning amputation serve as the sole determinants of their well-being. This circular pattern can result in a situation where the relentless pursuit of amputation is the only lens through which individuals with BID think they can reach an overall state of well-being. They may become fixated on the idea that amputation is the sole solution to their distress, potentially leading to the exclusion of alternative methods of coping or treatment. This perspective may also limit the capacity to objectively evaluate amputation decisions' long-term implications and consequences and the potential for alternative interventions or therapies. Even if that is not the case, and hedonism contemplates the trade-off between the immediate types of pleasures and long-term ones, the question remains whether such a perspective would still apply to BID.

> When we look backwards, we could afford to be selective. We ought to remember some of the bad events in our lives, when this would help us to avoid repetitions. But we could allow ourselves to forget most of the bad things that have happened, while preserving by rehearsing all of our memories of the good things. It would be bad for us if we were so selective when we are looking forward. Unless we think of all the bad things that are at all likely to happen, we lose our chance of preventing them. Since we ought not to be selective when

> looking forward, but could afford to be when looking backwards, the latter would be, on the whole, more enjoyable. (Parfit, 1986, p. 233)

Suppose an individual underwent many minor types of amputations, and each time, such an individual obtained pleasure from amputating a different body part, but only for a short period before plunging back into a state of suffering. In such a hypothetical situation, the short-term pleasure would necessarily result in long-term consequences which will be unbearable for the individual as he or she would reach a point in which not only will there be nothing else to amputate, hence preventing further pleasurable experiences resulting from such practices, but also the result of such practices will severely hinder one's ability to function both physiologically and psychologically. In such a situation, amputation may be seen as addictive, as can be the case with heroin for the junkie or with liquor for the alcoholic. Such a situation, however, can hardly be lifted and shifted to the BID individual, as the sole presence of other types of amputations on the BID body can lead the individual to fall outside the medical spectrum of the condition, as such type of parallel amputations are not pondered today in its officially recognised medical definition.

8.3.2 Arguments Against Hedonism

Pursuing pleasure and avoiding negative stimuli might lead the individual to be self-absorbed, thus preventing opportunities for self-growth. Similarly, such a chase of pleasure might lead to neglect of responsibilities and obligations, thus ignoring the broader sociocultural context within which the individual is entangled. Alternatively, considering pleasure the only element to achieve a well-being state can lead to also considering immoral types of pleasures, as might be the case for someone enjoying the suffering of others. From here stems another issue: The hard-line separating different types of pleasures. Such opposition goes that, given that Hedonism focuses on the pursuit of pleasure, little or no difference might exist between human existence and that of an animal, for example.

> If all that matters to my well-being is enjoyable experience, must there not come a point at which the value of the oyster life outweighs that of the life of Haydn. (Crisp, 2006, p. 631)

In Crisp's example, Haydn represents a composer who will achieve great things in life. Nevertheless, the oyster is used to highlight how the pleasure of the composer might very well be matched by the oyster's pleasures, assuming that this will live long enough and assuming the number of pleasures is the only criterion to evaluate one's life. The discriminating factor here is not the quantity of pleasure experienced but rather their quality. This opens the debate on what elements make one type of pleasure better than the other (see, for example, Feldman, 2004).

8.3.3 BID and Hedonism

Despite such compelling argumentation against Hedonism, it is easy to understand how these objections can only be considered valid by first anchoring them to a practical example. Assuming an individual lives a monastic life without any social relationship or social dependency, the hedonic perspective can make perfect sense. A similar case can be for individuals who disregard the social consequences of their actions and do not strive for positive social relationships. Such individuals can hardly be considered functioning members of society. Given that the present disambiguation on well-being aims to provide a framework for a better understanding of the BID phenomenon, such examples can be excluded from the reasoning as they would not add value to the discussion. BID individuals do not wish isolation. Similarly, amputation is seen as a way to enhance oneself. Assuming pleasure as the key to interpreting BID would not suffice nor provide any strengthening argument to justify the amputation request. Although the limb creates discomfort, amputation would presumably remove such discomfort and create well-being. What needs to be added is the consideration of the surrounding environment's perspective. Assuming this latter is also taken into consideration from a utilitarian perspective, that is *"a particular action is right if, and only if, in the situation, there was nothing the agent could have done instead such that, had the agent done it, the world, on the whole, would have been better"* (Tännsjö, 1998, p. 64), the question remains: how amputation would benefit the broader surrounding and, therefore, whether it should be permitted or not? Defending such a position might pose a significant challenge because it could be argued that permitting amputations would place an added strain on the healthcare system, a burden not currently experienced. This, in turn, could result in outcomes opposite to those advocated for by a utilitarian perspective. Such an argument might not make any more of a good point than advocating for free heroin to satisfy a drug addict's need for pleasure. If the surrounding environment is excluded, however, three problems remain: How to assess that such pleasure obtained through amputation is not a temporary amusement anticipating a greater dissatisfaction, as speculated in the example of the individual addicted to amputation? Similarly, how to assess whether amputation is indeed the right path to reach such a well-being state, that is, how to be sure the individual is not setting false expectations? Third, if there are no qualitative differences in types of pleasures, cannot the pleasure expected to be reached through amputation be a trade-off with another type of pleasure? Although literature seems to sustain the positive effects of amputation for BID individuals (Bayne & Levy, 2005, pp. 80–81; Blom et al., 2012, 2014), it would be far-fetched to sustain that amputation is indeed the best course of action for all individuals presenting BID symptomatology. A well-being theory liable to support the BID desire for amputation (that is, a convincing argument on the goodness of the amputation) must go beyond speculations about the everlasting pleasure of individuals. Demonstrating, rationalising and advocating for such well-being is necessary.

Individuals can be mistaken about what contributes to their well-being. The inherent subjectivity of the hedonistic approach opens the door to the possibility of misjudgement or error in assessing one's well-being. Individuals may believe they are experiencing well-being or pleasure, but external observations might suggest otherwise. The fact that individuals can misinterpret or misjudge their well-being does not necessarily render subjective theories entirely false. Although a good driving factor, pleasure is not a good enough reason as it seems to oversimplify the complexities of life. Notably, it overlooks the potential for growth and personal development that often arises from enduring hardships and making sacrifices. To illustrate this limitation, suppose that if Hedonism was the ultimate approach, any pleasure derived from amputation could theoretically be substituted by other forms of pleasure, provided there is no qualitative distinction. This notion alone challenges the idea that amputation for BID can be justified as the right course of action. Similarly, focusing on pleasure seems to move back BID on the same line as paraphilias or any disorders involving amputation to reach a hypothetical type of pleasure, whether this might be sexual or else. One possible way out of such an impasse would be to differentiate types of pleasures based on their quality so that a particular qualitatively higher kind of pleasure might equate to several qualitatively lower kinds of pleasure. For example, John Stuart Mill (2009) postulates the existence of qualitatively different types of pleasure, with mental pleasure outweighing the bodily type of pleasure.

> When, therefore, those feelings and judgment declare the pleasures derived from the higher faculties to be preferable in kind, apart from the question of intensity, to those of which the animal nature, disjoined from the higher faculties, is susceptible, they are entitled on this subject to the same regard. (Mill, 2009, p. 21)

Within such a paradigm, the BID kind of pleasure reached through amputation would not only need to be qualitatively different but also qualitatively superior to the type of pleasure an individual affected by paraphilia would reach through the same amputation procedure, given the fact that sexual pleasure might be seen as a lower type of pleasure. Adopting such a view might be particularly advantageous for BID, as amputation could be endorsed based on two premises: Firstly, the pleasure derived from amputation is qualitatively superior to any alternative pleasure that individuals with such desires might attain through other practices. This hinges on the idea that the desire for amputation results from a carefully considered judgment by the individual. The second premise follows the notion that, upon examining individuals who have undergone amputation, the pleasure derived from such a practice has been confirmed to be notably superior to other types of pleasures (Mill, 2009, pp. 16–17). This confirmation might lead individuals, if hypothetically placed back in time, to choose the same course of action and achieve the same outcome; having the amputation performed. However, this latter premise would necessitate relying on empirical evidence, which is currently limited for BID. Despite this possible way out, the underlying issue remains: how the individual can assess the type of pleasure that one ought to obtain from undergoing surgery; in other words, the problem of mistaken judgment or expectation on the right course of action is not extinguished even by taking into account qualitative elements of pleasure. However, a possible way around such a hindrance exists, which is to move the focus from pleasure to pain avoidance.

8.3.4 From Pleasure to Pain

Up to now, the discourse on Hedonism has gravitated around pleasure. The argument proposed is that by undergoing amputation, the BID individual would (ideally) reach such a state, thus achieving well-being. It is possible to utilise Hedonism from another perspective, which does not focus so much on pleasurable experiences but instead on the avoidance or removal of unpleasurable or painful experiences.

> Epicurus thought that avoiding pain including painful thoughts and regrets was much more important than seeking pleasure. (Binmore, 2020, p. 8)

Without indulging in Epicureanism, it is possible to use such an incipit to try switching the paradigm by considering avoiding pain as the main driving factor for BID and see how such a new approach would apply to some of the main opposition identified previously.

Taking pleasure as a central element within Hedonism raises three concerns regarding the potential benefits of amputation. Firstly, how does BID amputation contribute to the broader sociocultural context? Such a query considers the societal implications of the potential pleasure derived from amputation, highlighting the need to discern its impact beyond the individual. While it is not challenging to sustain, that amputation would benefit the pool of BID individuals, a more complicated task is to assert that granting such a procedure would somehow benefit individuals unaffected by the condition. The validity of such an accusation is easily nullified if the focus of the reasoning is switched towards pain.

Suppose an individual is affected by an unknown illness. Such an individual must take a particular type of medication to maintain a reasonable state of well-being and not plunge into unbearable suffering. Such a particular medication is supplied by the health care provider at a minimum cost to the individual. It is paid with taxpayer money since the governing institutions recognise the illness that affects the individual as debilitating. It is understandable how such a situation could burden the health care provider, assuming such medication needs to be provided promptly and that the number of people who request it is significant. A much better situation would be to have an alternate solution to such a particular type of medication. A solution that would cost less than the original medication and would need to be taken only once. Even better would be if such an alternative would bring a much sounder well-being state than the original medication because the individual taking it knows now that one will not have to repeat the process in the future, thus freeing one from such a specific burden. Suppose now that the justification for requesting such a medication would not be to avoid a state of debilitating suffering; instead, it would be to experience a particular type of pleasure. Within such an example, it is reasonable to understand how the fact of avoiding debilitating suffering would make a stronger argument than experiencing pleasure. Similarly, the existence of an alternative to the particular type of costly medication would be ideal, as it would present less of a burden for taxpayers. Such an example can be applied to BID. Supporting BID on the premises that amputation would bring pleasure seems a weaker argument compared to the argument that amputation not only would remove that specific kind of suffering (dysphoria) but also would be less of a burden on the healthcare system. This is because amputation is the only type of approach which has been successful, as of today, for BID individuals. This is to say that amputation might be the most advantageous practice, even from a cost-effective perspective, compared to long-lasting psychological and pharmacological approaches. Not only that, one might even sustain that by removing such suffering for the BID individuals, the latter will enhance the overall well-being of their sociocultural context by contributing in a productive way to its development. Such an approach to pain avoidance would also address the other two hindrances previously identified when talking about pleasure, namely the potential issue of temporary pleasure obtained through amputation and the problem of false expectations. While the temporary pleasure issue can be disqualified a priori from the pain avoidance reasoning, the problem of false expectation needs at least a mention. Amputation might not be the proper end to reach an overall state of well-being. However, from a BID perspective, it would help remove that type of pain that comes from having an alien limb similarly attached to one's body, as removing a cancerous appendix would do. Amputation within such a view is the proper action for BID.

8.4 Desire-Fulfilment Theory

The idea of desire is undoubtedly central to the BID debate. The Desire-Fulfilment theory proposes that an individual's well-being is defined by the extent to which one achieves desires or wants throughout time.

> What is good in itself for people and other subjects of welfare is their getting what they want, or the fulfilment of their desires, and what is bad in itself for them is their not getting what they want, or the frustration of their desires. (Heathwood, 2017, p. 135)

Such formulation implies that well-being is achieved by fulfilling as many desires as possible. Contrarily, dissatisfaction and frustration result from having desires not being fulfilled. Particularly relevant is the correlation between the strengths of the desire and the satisfaction of fulfilling it. The stronger the desire, the more satisfaction one would get by fulfilling it. A similar one-to-one correspondence exists for frustration. Within the Desire-Fulfilment paradigm, well-being can be framed in two distinct but interconnected ways: an aggregate or an overall perspective. Assessing an individual's well-being involves balancing desire's satisfaction and frustration throughout life. The first approach postulates that well-being is directly linked to the net outcome of fulfilling desires versus facing unfulfilled desires or frustrations. Therefore, a life with more fulfilled desires is generally considered better, while a life marked by significant unfulfilled desires and frustration is seen as worse. In this interpretation, well-being is a cumulative paradigm. Well-being can also be defined as the level of satisfaction an individual derives from one's life as a whole. In this case, well-being is evaluated by how much an individual's overall life goals and aspirations are met. It emphasises a comprehensive assessment of one's life. Singularly, a desire can be fulfilled even though the individual is not pleased, which differentiates this approach from Hedonism. The fulfilment of a desire does not always entail the experience of fulfilment, and specific desires can be satisfied without the individual having any conscious awareness or experiencing any specific conscious episodes during the fulfilment process.

Such an approach seems particularly fitting for the BID, given the direct correlation between desire fulfilment and well-being. Similarly, such an approach seems quite convincing as the main drive of action. One acts in a way that fulfils one's desire. It is not reasonable to think otherwise.

> What is intrinsically valuable for a person must have a connection with what he would find in some degree compelling or attractive, at least if he were rational and aware. (Railton, 1986, p. 9)

Despite such a compelling narrative, which naturally flows from every human action, there are some difficulties in such an approach.

8.4.1 Arguments Against Desire-Fulfilment

The first of such difficulties is categorising desires that are fulfilling in themselves and desires that are instrumental to something else. Suppose now, the BID individual desires amputation for the sake of it. That situation would not make much sense and could hardly be used to sustain any medical intervention. Suppose, alternatively, that the desire for amputation is instrumental to something else, such as the desire to end a state of suffering. While the BID individual would not accomplish anything in the first instance, in the second case, there would be a clear advantage that would be justifiable from an external perspective. Suppose now that the BID individual fulfils the amputation desire, which does not result in the satisfaction one was expecting to achieve. How can such two levels be framed within the Desire-Fulfilment theory? The underlying issue, therefore, is the existence of skewed desires, in other words, desires which were mistaken as fulfilling by the agent but are not as such once accomplished. In other words, an individual may believe that fulfilling specific desires will lead to a particular outcome or result. However, such desires, once achieved, result in something different. With this paradigm, it would be hard to distinguish which desires are worthy of contributing to well-being and which are not. Such formulation also leads to something else: the implication that such desire might not benefit the person but could even damage the individual. On-demand amputations for BID are a paramount example of such a dynamic. One might want amputation because it does not have all the facts laid out. If one knew that amputation would not accomplish the expected outcome, then very likely such desire would vanish or, at the very least, be categorised as irrational, thus disqualifying the individual from having medical decision-making capacity and, therefore, being recognised as an autonomous agent. Not all of an individual's desires additionally can be considered to impact well-being. Desires driven by confusion, ignorance or flawed reasoning can lead individuals to pursue goals or objectives that, upon closer examination, do not align with their best interests. These desires might not hold up to rational evaluation or reflection. Leaving such judgment to the individual allows for errors, as individuals might mistakenly assess the importance of specific desires to overall well-being. Similarly, it would be hard to comment, from an external observer, whether the desire the individual is after is intrinsically good or instrumental to something else, given that even the individual might mistake these two layers.

Suppose a physician is presented with an individual desiring amputation to achieve well-being. Nothing would suggest to such a physician whether amputation and well-being overlap and depend on one another. Finally, there could also be malicious or ill desires. Assuming the Desire-Fulfilment theory is accurate, one might desire the most vicious things. Nevertheless, such a desire might be considered worthy if it contributes to well-being. Again, there might be the case of desires not originating from the individual but instead considered obsessions. Such desires might hardly fall within a well-being ideology.

8.4.2 Informed Desire

Excluding the idea of BID being subjected to malicious or ill desires (which would disqualify the individual's reasoning a priori and make one worthy of blame, as discussed when addressing Moral Responsibility), what is left to consider is whether the absence of cognitive error or ignorance from a BID perspective would justify amputation.

> An individual's good consists in what he would want himself to want, or to pursue, were he to contemplate his present situation from a standpoint fully and vividly informed about himself and his circumstances, and entirely free of cognitive error or lapses of instrumental rationality. (Railton, 1986, p. 16)

According to this view, an individual's well-being is not determined solely by their desires or preferences but rather by what an idealised and fully informed version of that individual would want for one's good. Suppose such an ideal-advisor, for example, would recommend that pursuing limb amputation within this paradigm is intrinsically good for that individual's well-being. In that case, limb amputation is the path to choose. Although such a situation seems ideal, it is not immune from criticism. One possible criticism speculates the possibility that the individual's ideal-advisor's psychological makeup may evolve to such an extent that the desires they recommend for the original individual's well-being might not be any more appropriate for that individual (Rosati, 1995, pp. 307–313). Such an assumption relies on the postulate that the individual psychological constitution would significantly change once the role of the ideal-advisor is assumed. Such a change can be seen in two ways; such an advisor is still psychologically close to the original individual to provide a slightly better-informed guess about what is best for the individual. In such cases, the benefit of adopting such a view can be minimal. Alternatively, such an ideal-advisor reached a level of knowledge that shifted one's position to be distant from those of the original individual. In such a situation, the ideal-advisor might be seen as different from the original individual. A further position might sustain the ideal knowledge of the ideal-advisor, who knows what is best for the original individual. Such a situation would pose the ideal-advisor close to being omniscient. Such a situation seems to overflow towards transcendental debates. Moreover, its applicability is dubious. These and further oppositions are raised against a desire-centred approach to well-being. Nevertheless, it would be hard to go through them individually and sustain that they all apply to the BID case. Amputation might be the appropriate path for an individual as long as it is demonstrated that such a request comes from a critical assessment of the individual's condition. Ideally, such an assessment would need to dissipate any reasonable doubt about the individual's motivation and expectations.

8.4.3 Gender and Desires

A necessary step to explore the Desire-Fulfilment theory further is to apply it to practical examples. This need arises from the necessity to understand the potential benefits of fulfilling desires, the associated margin of error and the potential consequences of mistakes in judgments. Such an elaboration aims to formulate an educated assessment on whether amputation, for BID, is indeed the best course of action under this paradigm. Such reasoning also serves to answer one of the potential objections raised previously, that is, how a physician presented with an individual desiring amputation would infer whether amputation and well-being overlap and are dependent on one from the other. While the data available today regarding BID amputation is still partial, it shows promising results regarding post-amputation satisfaction. However, it is necessary to examine cases that are better established within the medical field than BID. Doing so makes it possible to draw meaningful conclusions about applying the Desire-Fulfilment theory.

SRS is one such case frequently referenced throughout the present discussion to illustrate the challenges of body-modifying surgeries and highlight the unique aspects of BID. Individuals who experience distress and relative impairment in functionality due to a misalignment between gender identity and assigned sex at birth may very well fall within the spectrum of Gender Incongruence or Gender Dysphoria, depending on the taxonomy at hand. Such a diagnosis may serve as a trigger to seek out medical and surgical interventions. These interventions can encompass a range of gender-affirming procedures and therapies, such as hormone replacement therapy (HRT), gender confirmation (or reassignment) surgeries and various forms of psychotherapy. The goal of these interventions is to alleviate the state of dysphoria originating from the incongruence between one's gender identity and physical characteristics. The decision to pursue these medical and surgical interventions is not made lightly. It typically involves in-depth discussions with healthcare professionals, including mental health specialists and medical experts who specialise in this specific branch of healthcare. These consultations help individuals make informed choices about the best course of action. From such complex types of procedures, chiefly SRS, and similarly complex assessment procedures, one would expect a high satisfaction rate for individuals who successfully transitioned, but this does not always seem to be the case, with noteworthy suicide rates reported among the transgender communities (Wiepjes et al., 2020) as well as individual de-transitioning (Richards & Doyle, 2019) with this latter term referring to "the process of reidentifying with one's birth sex after having undergone a gender transition" (Expósito-Campos, 2021, p. 270) to the extent to call for a reassessment of the criteria used today for grant SRS and similar type of requests (for an overview see for example Entwistle, 2021). Two problems, therefore, seem to be present today within the SRS debate; the need to refine further the criteria to undergo such procedures to limit false positives (that is, giving the go-ahead for surgery to individuals who might need other types of intervention) and assessing the needs of those who decide to de-transition (Hildebrand-Chupp, 2020; Vandenbussche, 2022). Such a need is compounded by a spike in Gender Dysphoria cases reported since an early age, for which no apparent reason is currently known (Aitken et al., 2015; Kaltiala et al., 2020), which considered together with the previously discussed trends concerning suicide rates and de-transitioning practices cannot but raise certain concerns. This assessment appears to reintroduce challenges in effectively addressing psychological distress states that lack clear physiological roots, particularly when considering irreversible medical interventions. The complexity arises from the fact that psychological distress often involves intricate factors that are not easily linked to physiological markers. This is not only because the originating desire to transition (or to amputate, in the BID case) might be wrong a priori, as it might be the case for a skewed type of desire or the desire resulting from wrong emotional or cognitive elaboration. This might also be because desires change through time, and what once might have been considered necessary could leave room for other priorities. This is to say that more impending desires might overwhelm the initial desire, making it not as important as initially thought (for a case study on the reasons for de-transitioning, see, for example, the survey of Littman, 2021). By reflecting on the drawbacks of SRS surgery, it is possible to reflect on two intrinsic issues that might hinder the BID position; the issue of re-prioritisation of desire through time and the fact that desires are directed towards a future state of affairs.

The first issue is the re-prioritisation of desires. Suppose an individual desires to be of the opposite sex and gender, and to do that, one needs necessarily to go through SRS. Suppose now that such an individual was right (hence, no skewed desires nor misjudgements are in place) and that SRS and the change of sexual characteristics lead to the well-being they seek. Suppose now, again, that shortly after the surgery, such an individual realises that the new state is, for whatever reason, hindering one's social relationships, which suddenly become more important to maintain the state of well-being than the newly acquired sexual and gender-specific characteristics (social stigma is for example, among the causes identified by Littman for de-transitioning). Within such a situation, although SRS seem to be the right path, because indeed it brought the expected outcome, it also brought along the awareness that there are also hidden elements (namely social relationships as for the example) which were taken for granted up to that very moment, and which suddenly were taken away. One might wonder at this point whether SRS surgery was not an instrumental value for something else (new sexual characteristics), which clearly it was, but also whether the desire to have different sexual characteristics was instrumental to something else (namely reaching a general idea of well-being). Such a reasoning can be applied to any state. The issue at hand, therefore, seems to be how to assess the rightfulness of the desire chosen, granted the fact that all desires can be instrumental to something else and that, given such nature (namely being instrumental), can very well change through time and space. One might desire something in the present, but upon obtaining it, realise they were better off without it, leading to a shift in desire towards something else.

This brings into the picture the second issue; desires are directed towards a future state of affairs, which resides in a speculative realm. One cannot be sure that a desire, once obtained, will bring any added value. In fact, one can desire to be worse off compared to a present condition, and according to the Desire-Fulfilment theory, such desire would still contribute to one's overall well-being.

8.4.4 *Amputation as an Ideal Desire*

Once some of the core points of the Desire-Fulfilment theory are presented and highlighted, applying such a theory to BID and assessing whether amputation is suitable for the individual is possible. Similarly, it is possible to speculate on the reasonable criteria that the physician presented with an on-demand type of amputation request from a BID subject needs to take into account.

Suppose one wishes to have his right leg removed to restore a sense of completeness with the Self and remove the state of dysphoria. In this case, amputation is instrumental to something else; a sense of completeness. The very first problem in such a situation is that there is no guarantee that amputation will have the expected outcome based on the previously moved opposition that desires reside in a speculative realm projected into the future. That is to say; there is no way for the individual to base one's reasoning on a future and never-experienced state of affairs. Similarly, it would be hard to sustain that amputation, namely a medical practice, is the only way to reach such a state. This is based again on speculation, which may lead to countering such reasoning with arguments advocating alternative medical interventions. Since the individual desires a future state of affairs, one might argue that such medical aid, before being released, must satisfy first prerequisites other than the individual's desire. Based on the above elaboration, two potential issues must be addressed before answering whether amputation is the right path for BID according to a Desire-Fulfilment perspective. It is possible to retrieve some concepts that emerged during the fieldwork phase and some theoretical elaboration articulated in the preceding chapters. Firstly, the issue of instrumental desire. It is clear that amputation is not an end in itself; instead, it is an instrumental type of desire. Within such a view, achieving amputation per se would not add value to the individual. Such an assessment is valid. If an alternative to amputation exists, then this latter would cease to have its relevance, but since such an alternative currently does not exist, logic dictates that such instrumental desire (the desire to have a limb amputated) is a necessary part of achieving the ultimate end, that is a state of well-being. If there were a magic pill that one would need to take to reach the desired state immediately, this would likely be the preferred pathway. Such an alternative, however, does not exist. This leads directly to the second issue. It is impossible to predict with certainty that such an instrumental desire would indeed lead to the desired state since such a desired state resides in a future never experienced by the individual before. Again, such a reasoning is correct. Still, its correctness can hardly be used to sustain that amputation is wrong. Similar things can be said for any medical interventions, as amply debated in the preceding chapters. What is needed from the physician's end is a way to dissipate reasonable doubt. This is done with what previously were identified as core criteria for any medical procedure, namely emotional stability and lack of impulsivity. Suppose such criteria are satisfied and complemented by particular reasoning on why such a specific request is needed from the patient's perspective (the trigger point of action that leads to exiting the current status quo). In that case, assuming that the individual is rational is possible. Similarly, one's desire is to be considered informed within reasonable expectations.

Although such a view would likely suffice to justify the BID desire as rational, one can strengthen the argument further by retrieving another element addressed throughout the present work and first presented when introducing the concept of autonomy. Such an element is the idea that one is truly free (hence autonomous) if and only if one can shape and govern one's desires. This implies a two-tiered structure of desires: First-order desires refer to immediate wants and inclinations. In contrast, second-order desires involve a reflective layer of desires concerning one's first-order desires (Frankfurt, 1971). Within such a view, autonomy (and therefore rationality) is manifested not only in the capacity to act according to one's desires but also in the higher-order ability to have control over the very desires that motivate actions; in other words, one is truly an autonomous agent if he or she possesses the ability to desire what one want to desire and to shape action accordingly. Within such a view, a BID individual would not be autonomous if the desire to amputate would be a first-order desire, and despite that, one would not have the control to oppose such a desire. This would be different, however, if the desire to amputate would be aligned with the desire to desire to amputate (second-order). In this last instance, the amputation would result from a pondered decision. Now, it will not be the scope of this section to dwell any longer on the rationality of the BID desire, as it has been done plenty of times already. Instead, retrieving such ideas, which have been used previously through the present work, was deemed necessary to highlight how a Desire-Fulfilment account of BID would align with the previously presented reasoning. Following such line of reasoning, amputation would be a good way forward for BID.

8.5 Objective-List Theories

Objective approaches imply the existence of elements that are inherently good and independent from the reactive attitude that the individual holds towards them (Rice, 2013). In other words, certain elements are considered good or bad without the individuals necessarily recognising them as such.

> Certain things are good or bad for people, whether or not these people would want to have the good things, or to avoid the bad things. (Parfit, 1986, p. 499)

Such characteristics are generally shared and agreed-upon elements across individuals. For example, Carol Diane Ryff (1989) identifies several elements, such as the ability to make choices and decisions independently, exert control over one's own life and make meaningful and positive connections with others, such as family, friends and communities, contributing to a sense of well-being. It follows the idea of having a purpose in life, which gives the individual a sense of direction and personal growth, namely striving for self-improvement and development, seeking to become the best version of oneself. It would be hard to defend the position that what Carol Diane Ryff postulates, for example, are not elements widely accepted. Everyday judgments and choices regarding well-being provide, to some extent, a compelling and defeasible explanation for adopting Objective-list theories (Fletcher, 2017, p. 152).

8.5.1 Challenges with Attitude-Independence Approaches: Human Nature

One of the main problems with such approaches is exactly their pluralist status; in other words, who decides what are objective values of well-being and why is that?

> Pluralism seems objectionably arbitrary. Whatever the composition of the list, we can always ask: why should these things be on the list? What do they have in common? What is the rational principle that yields the result that these things, and no others, are the things that are good? (Bradley, 2009, p. 16)

A straightforward answer to that and similar questions might be that items in such lists enhance well-being. To avoid circular reasoning, a deeper examination of the common characteristics, values or attributes that underlie the diverse elements listed is required.

> Pluralists must tell us, for example, how to compare the effects on well-being of a certain amount of pleasure with the effect of a certain amount of knowledge . . . To the extent that the pluralist refuses to tackle these questions, she abandons the philosophical project of understanding well-being; she admits defeat. A theory that tells us that A, B, and C are intrinsically good, but does not tell us why those things are on the list or how to weight them, does not give what we initially wanted out of a theory of well-being. We wanted enlightenment, but we are provided instead with a list and told not to look any deeper. This is not theorising, but a refusal to theorize. (Bradley, 2009, p. 16)

One way to reconcile this opposition is by postulating intrinsic goodness in certain elements. All individuals possess inherent capacities that are ingrained in human nature. These unique attributes define the human species and are fundamental to human existence. The nurturing and development of such abilities and attributes, therefore, is what brings well-being. Such an approach would ideally satisfy the opposition to the arbitrariness of what goes on the list by identifying the commonality of human nature (Bradford, 2017). Such a view opens the path to a whole different dilemma on what it entails to be human. Hurka (Hurka, 2009), for example, links such characterisation to the capacities that hold particular relevance within the context of human existence. Humans are naturally gifted with certain distinctive abilities and characteristics necessary for survival and growth. Although such an account provides a better overview of what could be included in an objective list of well-being, it still leaves open the question of how to pinpoint such elements. In other words, what methods and methodologies are used to identify such criteria, and are such criteria applicable cross-culturally? Assuming such issues were absent (namely, having a clear and concise view of what it means to be human), another question arises; whether it is enough to possess such elements or whether well-being is dependable also on the way such elements are exploited. The debate on human nature seems particularly fitting for the BID case, as it was previously mentioned that one characteristic identified for the rejection of amputation is the intrinsic idea that such practice would alter a given way of existing, that is, body harmony. Whether this is a valid point or not needs to be assessed once, and if there ever will be a method to clearly identify what renders humanity what it is today, whether any divergence from such hypothetical norm is deleterious. Until then, it is hard to comment on what would be the perfect elements of human existence and whether the BID individual would alter any of them by undergoing amputation.

8.5.2 Alienation and Health-Related Matters

A more compelling objection to the Objective-list theory of well-being is the concern that such theories might not adequately account for individual subjectivity, potentially resulting in a conception of well-being that feels alienating or disconnected from people's daily experiences (Hall & Tiberius, 2017, pp. 182–184). In essence, a well-being concept may face challenges if it allows for the possibility that a person could theoretically possess a high level of well-being (namely, the individual possesses all elements listed as bringing well-being), even in a scenario where such an individual remains untouched by such possession, in the sense of not experiencing any pleasure, drive or cognitive engagement (Fletcher, 2017, p. 156).

An objective approach to well-being seems hard to defend for BID, given the uncertain aetiology of the condition and its treatments. Suppose positive social relationships are an understandable objective goal to reach well-being. The same thing cannot be said for health, for example, due to the complexity of the concept influenced by various subjective and objective factors. It is important to note that such a concept cannot be determined solely by measurable physical parameters. An individual's perception and experience also influence it. One person might consider something healthier than another individual. Suppose a certain individual considers smoking the healthier choice among the ones at his disposal, such as binge eating and taking drugs. Assuming such elements within one's life are mutually exclusive, one could make the conscious choice that among all of them, smoking is the least hazardous activity. For instance, someone with a chronic illness may have a different health perception than someone without health issues. In the first instance, the individual will have a different view of what is healthy and healthier tailored to specific instances, which likely can be used as a benchmark and was built on life experiences of dealing with such a condition. In the second case, such a benchmark will be fruitless as the individual without a chronic illness will find it very likely restrictive and without purpose. Similarly, cultural and societal variation, personal values and goals, and biological makeup can all impact one's understanding and perception of health. For example, the definition of mental health and well-being has evolved over the years, reflecting changing societal attitudes and scientific discoveries. For instance, Sumner (Sumner, 1999) highlights that adopting an objectivist perspective cannot elucidate why certain factors are beneficial for that person.

> Since an account of the nature of welfare is descriptively adequate only if it is faithful to our ordinary concept, any serious contender must at least preserve the subject-relativity which is definitive of prudential evaluation. It cannot manage this much then, though it might be a plausible rendering of some other dimension of value, it is not a theory about welfare at all ... it is not too much to expect a theory to explicate this feature, to tell us what it means for my life to be going well not just in itself or from some other standpoint but for me, to explain how it is that lives can have this peculiar perspective of value. (Sumner, 1999, p. 20)

Similarly, Connie Rosati (2020) put forward a rational fit theory of well-being, in which the elements that contribute to an individual's well-being must be aligned explicitly with that person's unique and individual nature (as mentioned, for example, by Hall & Tiberius, 2017, pp. 175–176). Such remarks serve only to highlight some of the problems with an attitude-independent approach.

Such a take, however, does not necessarily preclude the validity of Objective-list theories. For example, John Finnis (Finnis, 2004) argues that maximising all elements on a well-being list is futile. According to his perspective, individuals can selectively embrace specific elements recognised as basic goods for human life. Although such framing renders objective accounts of well-being quite flexible, it seems at the same time to fall back on some of the issues stated previously. If one starts to cherry-pick which account of Objective-list theory is the best, one might as well start integrating these letters within a subjective stance of well-being. The existence of so many declinations of such a theory seems to defy per se the claim of being attitude-independent, with this latter being not so much so for the individual who is subjected to it but rather on the author who formulates it. Taking the framing of Finnis further once again, it is possible to speculate on the meaning of such basic values. Take, for example, the value of life.

> Life here signifies every aspect of the vitality which puts a human being in good shape for self-determination. Hence, life here includes bodily (including cerebral) health, and freedom from the pain that betokens organic malfunctioning or injury. (Finnis, 2004, p. 224)

Even though there is no doubt that the idea of health must have a solid foundation shared cross-culturally, it is equally true to recognise that simply stating health as a characteristic without dwelling on its nuances seems to leave too much room for interpretation. If that is the case, one might wonder why there is a need to entrust the concept's meaning to an external party and if it would not be better to leave it to the subject upon which health-related issues are being debated.

8.5.3 An Objective Take on Health

The concept of health has been regarded up to this point as a fundamental component of well-being without providing any definition. In its most strict sense, health implies a state of proper functioning, encompassing an organism's physiological and psychological aspects. In its broader sense Health is a state of complete physical, mental and social well-being and not merely the absence of disease or infirmity (World Health Organization, 2020, p. 1). The definition provided by the WHO for the present scope seems circular: Health is well-being, and well-being is health. For this reason, a more strict definition of health will be adopted for now. From the empirical phase of the research, this latter definition emerged as a rule of thumb. In this fundamental sense, health implies that the organism operates optimally. Conversely, a decline in health can take the form of injury, illness, disease or impairment and manifest when a part of the organism malfunctions in a way that is, in some relevant sense, inferior to the expected standard (Schroeder, 2017, pp. 222–223).

Health is not only perceived as a critical element of well-being, but it can also be linked to the innate drive for self-preservation (Finnis, 2011). Moreover, the connection between health and well-being is also evident in the impact of impaired health on an individual's ability to flourish. Flourishing refers to realising one's full potential and attaining a fulfilling and meaningful life. When health is compromised due to physical or mental conditions, it can significantly hinder an individual's ability to reach one's full potential.

> If you say of a human being that he is flourishing, your statement is thrown into doubt if he is correctly described either as psychologically or as physically unhealthy, weak, damaged, and stunted. Certainly, if he suffers from both kinds of disabilities, the claim that he is flourishing, or that he is doing well, is impossible to sustain … It is unobjectionable to say that everything that interferes with some component of a human being's flourishing, by making it more difficult or impossible for him to flourish, is bad for him. (Murphy, 2001, pp. 133–134)

However, such a stance on the relationship between health and well-being is not always accurate, with physical and psychological impairment affecting well-being perception far less than postulated. While good health can contribute to higher overall well-being, it is not its sole determinant as it can be accompanied by other elements such as resilience, social support, drive for self-improvement and similar (Angner et al., 2013; Riis et al., 2005; Ubel et al., 2005). Health constitutes a foundational element of overall well-being. In its absence, well-being is inevitably diminished, yet the level of well-being may remain relatively high if other crucial elements are adequately in place (Carel, 2009). Several approaches can be used to understand the correlation between well-being and impaired health. The first of such approaches draws from the concept of hedonic adaptation (Frederick & Loewenstein, 1999). According to this idea, people can adapt to unfavourable circumstances, including illness, over time. Hedonic adaptation suggests that, as time passes, individuals tend to experience a reduction in the intensity of their emotional responses to positive and negative stimuli. In the context of health challenges, this means that, despite the initial impact of an illness on one's well-being, individuals can gradually adapt to their new circumstances and find a way to maintain a level of well-being. This adaptation might involve a shift in their perspective, a focus on what they can still enjoy, or the development of coping strategies to mitigate the emotional impact of the illness. A further approach centres on the idea that individuals coping with illness can compensate by finding satisfaction in other areas of life (Lucas, 2007). Such individuals may derive a sense of well-being from domains outside of health, such as work, family life, hobbies or personal relationships. This capacity for compensatory well-being allows individuals to maintain a reasonable quality of life despite health challenges. Accepting that adaptation to illness or impairment has no limits and that all medical conditions or impairments can be equally well-adapted would be preposterous. It seems logical to assume there must be a point after which illness or impairment becomes too profound to be mitigated through adaptation or compensation. One such threshold might be the extent to which the illness or the impairment disrupts the individual's daily functioning. Medical conditions that significantly disrupt an individual's ability to carry out daily activities, pursue their goals and engage in meaningful social interactions may have a deleterious and long-lasting impact on one's well-being. Similarly, degenerative diseases deteriorate an individual's functional capacity, making adaptation more challenging as the condition progresses. Finally, the social stigma associated with a specific condition might also be deleterious to well-being (Clarke et al., 2011). If these three elements are considered valid—which does not necessarily mean that they are—what seems to be a cutting line to compensate for the well-being missing from physical or psychological impairment or illness is a time factor needed for adaptation. This means that the daily reiteration of the illness becomes ipso facto a disability, with all its sociopolitical acceptations (Fava & Sonino, 2008). When an illness evolves into a disability, it can impede the individual's capacity to adapt and counterbalance the physical or psychological challenges with other elements of one's life.

Such a short elaboration can be used to infer that health seems to possess an objective quality regardless of its definition. Despite how health is conceptualised, there appears to be an inherent standard below which life loses its appeal. In other words, an ideal threshold exists, after which sensible individuals may reasonably conclude that life is not worth it. This implies a universally recognisable standard, suggesting that specific objective criteria define the fundamental level of well-being below which the appeal of life is nullified.

8.5.4 Amputation as the Objective Good

Health is to functionality as illness or diseases is to impaired functionality. This means an individual with four functioning limbs and no psychological impairment is healthy. Conversely, if such an individual decides to have one of the limbs removed, such an individual would lose one's health status. The first issue with such an assumption is that reduced health that affects a capability that the individual would not have utilised anyway despite their health condition may not necessarily have a negative impact on that person's overall well-being (Schroeder, 2017, pp. 224–226). In such cases, the restriction may not translate into a practical or meaningful loss for the individual. Suppose an individual has a health condition that limits one's ability to engage in a specific physical activity, but such an individual had no prior interest or intention to participate in that activity; in this case, the health-related restriction may be negligible to one's well-being. In these situations, the restricted capacity is not a meaningful component of the individual's well-being, as it is not a part of one's life or goals. Such a position, however, would take much work to sustain for BID. In this latter case, it is feasible to speculate that one still uses one's legs or arms daily even though one might not want to do so, with some exceptions limited to a specific time window, as is the case for the pretending behaviour. Such argumentation might lead to sustaining that the individual speculates—based perhaps on wrong cognitive assumptions—that no loss of well-being will arise from losing a limb, or even it will enhance one's well-being as per standard BID narrative. Nevertheless, the actual realisation will severely impact one's well-being because of such wrongful speculation. Suppose a scenario in which a planned activity can no longer be done due to unforeseen circumstances. The impact on one's well-being will likely depend on the nature of the alternative chosen. If such a choice is inferior to the first, then it is feasible to speculate that there will be a decline in one's well-being. Conversely, if the second-choice activity turns out to be better, in some regards, than the initial choice, it will increase well-being. Such formulation seems more plausible for the BID as it leverages the enhanced psychological functionality of the individual, which ideally would lead to expressing one's full potential, which could not have been done otherwise. Despite the fitness of such formulation for BID, one problem needs to be solved. How can one be sure that the alternative to the current status quo—namely, keeping all four limbs—would bring added value to the BID individual? The challenge to overcome, therefore, is how to ascertain the nature of the alternatives. For the sake of argumentation, it is possible to speculate that amputation will bring a value more significant than the one the BID individual obtains in daily activities. Such added value resides not so much on the physical level but instead on the psychological domain. A further problem is how to frame instrumental types of activities within a less-than-ideal sociocultural context. In other words, assuming impairment for the BID individual would unlock one's full potential, how much will such an ideal full potential be, in turn, impacted by environmental challenges that are not dependent on the individual? Such environmental challenges do not automatically imply diminished well-being; on the contrary, they could have the opposite effect. The sense of fulfilment derived from conquering these adversities has the potential to significantly boost one's overall well-being through a sense of self-worth. These insights highlight that missing certain capabilities due to illness or impairment does not necessarily mean a decline in well-being. In some instances where capabilities are lacking, it is conceivable to envision equally valuable replacements for activities pursued by their intrinsic worth and equally effective methods for executing instrumental tasks. In specific situations, however, these alternatives may not be readily accessible. Furthermore, even when equally valuable replacements are within reach, there is no assurance that individuals will avail themselves of these alternatives (Schroeder, 2017, pp. 224–225). Hence, the degree to which the loss of capabilities due to ill health or similar impacts one's well-being hinges on the frequency with which individuals can genuinely identify equally valuable substitutes and equally efficient alternatives (Schroeder, 2017).

Despite such nuances, one thing seems clear: Health for BID individuals means psychological enhancement over physiological functionality. This is not to say that physiological functionality is to be discarded from a BID perspective; instead, physiological functionality still constitutes health in a different and perhaps subsidiary way compared to psychological functionality. This is to say that despite the interpretations used for its understanding, health retains an objective characteristic. Such objective characteristic is the hypothetical benchmark identified previously, below which life loses its appeal. This is to say, again, that there is an ideal threshold beyond which rational individuals may reasonably deem life not worth living. Such a threshold can be envisioned as the baseline of health, whether psychological, physical or else. Within such a view, health can be depicted as all those elements (psychological and physiological or whatever these might be) that the individual recognises as keeping one afloat above such hypothetical threshold. Once such elements are taken away or seriously hindered to the extent of one being unable to remain afloat, health is lost and life loses its value. It would ideally be a disadvantageous cost-reward ratio in which the cost of living outweighs its rewards. For BID individuals, the presence of a persistent state of dysphoria, a state which hinders one's ability to function both physiologically and psychologically in a sound way and which might very well become more persistent and overwhelming over time, might be seen as impacting health to the extent of making one sink below the hypothetical threshold. Within such a paradigm, losing a limb would not make one's health plunge into the abyss, as might be inferred per standard account. Instead, losing a limb through a voluntary and pondered action (amputation) might be seen as removing a weight dragging down the individual towards the bottom of the swamp. Amputation, in this case, could be seen as cutting the line that anchors the individual to the weight that eventually would bring one below the hypothetical health threshold.

It would be unreasonable to sustain, within such a view, that health is not an objective element. Similarly, it would be unreasonable to sustain that amputation is not the correct choice for the BID individual. It is better to cut lose any hindrances to one's health that prevent reaching well-being before waiting too long and, ideally, passing a point of no return. Suppose a castaway individual with a badly injured leg. The castaway individual might be forced to amputate their limb due to its gangrenous nature and save their life. Similarly, the BID individual might be forced to amputate their limb due to its alien nature and save their life. In one case, the reluctance to amputate might lead the gangrene to spread to other body parts and be deadly, while in the other, the dysphoria might overwhelm the rationality of the BID individual and push them to take extreme measures. In both cases, deciding not to amputate will seriously affect the individual's health. The difference between the castaway person, the BID individual and anyone else is whether the stress should be on the physiology, the psychology or something else. Independently where this may be, health within such a view is and remains an objective value.

8.6 Conclusion

The purpose of the chapter was to answer whether it is good for BID subjects to undergo amputation. Three theories were proposed to provide such an answer: Hedonism, Desire-Fulfilment and Objective-list theories. Hedonism asserts that whether something contributes to well-being always hinges on attitudes. One must exhibit some subjective response for an element to be considered good. Such an approach leaves room for individual interpretation. Applying Hedonism to BID, one might put forward the differentiation between higher and lower types of pleasure, with the BID type of pleasure as the former kind, derived from feelings and judgment as postulated by John Stuart Mill (2009). Such an approach includes a more reasonable quest to seek pleasure than the one obtained through purely emotional or instinctive pursuit. Similarly, there might be the case of grounding some qualitatively higher pleasure within a utilitarian framework.

> If I am asked, what I mean by difference of quality in pleasures, or what makes one pleasure more valuable than another, merely as a pleasure, except its being greater in amount, there is but one possible answer. Of two pleasures, if there be one to which all or almost all who have experience of both give a decided preference, irrespective of any feeling of moral obligation to prefer it, that is the more desirable pleasure. If one of the two is, by those who are competently acquainted with both, placed so far above the other that they prefer it, even though knowing it to be attended with a greater amount of discontent, and would not resign it for any quantity of the other pleasure which their nature is capable of, we are justified in ascribing to the preferred enjoyment a superiority in quality, so far outweighing quantity as to render it, in comparison, of small account. (Mill, 2009, pp. 16–17)

A Hedonist embracing such a view might see BID amputation desire and commend its goodness because the pleasure obtained through amputation is qualitatively higher than any other type of pleasure such individuals might obtain through other practices. Supporting amputation based on pleasure might nonetheless still be seen as not a strong enough argument. Instead, when the focus shifts from seeking pleasure to mitigating or eliminating the distress experienced by individuals with BID, new elements emerge. From a cost-effective perspective, amputation might be the most advantageous practice compared to prolonged psychological and pharmacological interventions. Furthermore, by alleviating the suffering of BID individuals through amputation, there is a potential enhancement of the overall well-being of the sociocultural context. This suggests that these individuals, freed from the burden of dysphoria, could contribute more productively to societal development. This approach to pain avoidance also addresses concerns related to temporary pleasure obtained through amputation and the issue of false expectations. Amputation may not be the definitive solution for achieving an overall state of well-being. However, from a BID perspective, it effectively eliminates the pain associated with having an alien limb attached to one's body. In this view, amputation emerges as a justified and appropriate action for individuals dealing with BID.

The second theory considered for the BID amputation's goodness was the Desire-Fulfilment theory, which posits that fulfilling one's desires is the key to well-being. However, this view becomes problematic when instrumental types of desires, desires derived from ill intentions or simply mistaken desires due to ignorance or similar circumstances come into the picture. To oppose such a problem, the existence of an ideal-advisor, namely a speculative entity who can comment on one's position from a much more knowledgeable perspective, was introduced. Although such an entity would likely address skewed desires, some problems remain, such as whether the ideal-advisor would apply to any practical example as it resides almost exclusively within the boundaries of a thought experiment. Despite such oppositions, it is problematic to sustain that there is no correlation between desire and well-being, and similarly, getting what one wants is not the trigger for any human action and, as a consequence, at the very least, a positive thing for the individual. By following such reasoning, it is easy to see how it would be hard to reject the idea that amputation is wrong for the individual since this is what one wants, even though it is an instrumental desire for something else, that is, well-being. The opposition of skewed desires, desires transforming through times (hence being unstable) and desires directed towards a future state of affairs, can be countered by drawing on practical examples; as of today, within medical practice, an individual's initiative is considered as a core element to infer one's willingness to change a current status quo in light of a future, better Self. Since it is impossible to predict the future state of affairs to justify the action, one needs to dissipate doubt within reasonable expectations. To do so, the criteria in use today are emotional stability and lack of impulsivity. If such criteria are satisfied, then BID amputation can also be envisioned as the right action for the individual. To further strengthen the justification for BID desires as rational, one can draw additionally from a two-tiered structure of desires: first-order desires pertain to immediate wants and inclinations, while second-order desires involve a reflective layer of desire concerning one's first-order desires. In this expanded perspective, autonomy is expressed through the higher-order ability to control the desires that drive actions.

Finally, what objectively serves a patient's best interests was considered by evaluating the broad umbrella of Objective-list theories. However, such approaches present specific challenges, particularly their pluralistic nature. A possible way around such a hindrance is to identify a common factor shared among all individuals indifferently. Such an elaboration was expanded by speculating that health seems to exhibit an objective quality irrespective of how one defines it. Regardless of the conceptualisation of health, there exists an inherent standard beneath which life loses its attractiveness. This implies the existence of an ideal threshold beyond which rational individuals might reasonably decide that life is not worth living. This perspective suggests a universally recognisable standard that transcends subjective interpretations of health and well-being, making such a concept easily applicable to BID.

In conclusion, Hedonism, Desire-Fulfilment theory and Objective-list theories are valid frameworks for addressing whether amputation is beneficial for individuals experiencing BID. Their adaptability allows them to be either allies or adversaries, depending on the specific aspects highlighted in the evaluation.

References

Aitken, M., Steensma, T. D., Blanchard, R., VanderLaan, D. P., Wood, H., Fuentes, A., Spegg, C., Wasserman, L., Ames, M., Fitzsimmons, C. L., Leef, J. H., Lishak, V., Reim, E., Takagi, A., Vinik, J., Wreford, J., Cohen-Kettenis, P. T., De Vries, A. L. C., Kreukels, B. P. C., & Zucker, K. J. (2015). Evidence for an altered sex ratio in clinic-referred adolescents with gender dysphoria. *The Journal of Sexual Medicine, 12*(3), 756–763. https://doi.org/10.1111/jsm.12817

Angner, E., Ghandhi, J., Williams Purvis, K., Amante, D., & Allison, J. (2013). Daily functioning, health status, and happiness in older adults. *Journal of Happiness Studies, 14*(5), 1563–1574. https://doi.org/10.1007/s10902-012-9395-6

Bayne, T., & Levy, N. (2005). Amputees by choice: Body integrity identity disorder and the ethics of amputation. *Journal of Applied Philosophy, 22*(1), 75–86. https://doi.org/10.1111/j.1468-5930.2005.00293.x

Binmore, K. (2020). *Crooked thinking or straight talk? Modernizing epicurean scientific philosophy*. Springer. https://doi.org/10.1007/978-3-030-39547-6

Blom, R. M., Hennekam, R. C., & Denys, D. (2012). Body integrity identity disorder. *PLoS One, 7*(4), e34702. https://doi.org/10.1371/journal.pone.0034702

Blom, R. M., Braam, A. W., de Boer-Kreeft, N., Sonnen, M. P., & a. M. (2014). Body integrity identity disorder, relief after amputation. *Tijdschrift voor Psychiatrie, 56*(1), 54–57.

Bradford, G. (2017). Perfectionism. In G. Fletcher (Ed.), *The Routledge handbook of philosophy of well-being* (pp. 124–134). Routledge.

Bradley, B. (2009). *Well-being and death.* Oxford University Press. https://doi.org/10.1093/acprof:oso/9780199557967.001.1

Carel, H. (2009). I am well, apart from the fact that I have cancer: Explaining well-being within illness. In L. Bortolotti (Ed.), *Philosophy and happiness* (pp. 82–99). Palgrave Macmillan.

Clarke, P. J., Ailshire, J. A., Nieuwenhuijsen, E. R., & De Kleijn-De Vrankrijker, M. W. (2011). Participation among adults with disability: The role of the urban environment. *Social Science & Medicine, 72*(10), 1674–1684. https://doi.org/10.1016/j.socscimed.2011.03.025

Crisp, R. (2006). Hedonism reconsidered. *Philosophy and Phenomenological Research, 73*(3), 619–645. https://doi.org/10.1111/j.1933-1592.2006.tb00551.x

Entwistle, K. (2021). Debate: Reality check – Detransitioner's testimonies require us to rethink gender dysphoria. *Child and Adolescent Mental Health, 26*(1), 15–16. https://doi.org/10.1111/camh.12380

Expósito-Campos, P. (2021). A typology of gender detransition and its implications for healthcare providers. *Journal of Sex & Marital Therapy, 47*(3), 270–280. https://doi.org/10.1080/0092623X.2020.1869126

Fava, G. A., & Sonino, N. (2008). The biopsychosocial model thirty years later. *Psychotherapy and Psychosomatics, 77*(1), 1–2. https://doi.org/10.1159/000110052

Feldman, F. (2004). The objects of enjoyment; mill's hedonism. In *Pleasure and the good life: Concerning the nature, varieties and plausibility of hedonism* (pp. 71–78). Clarendon Press.

Finnis, J. (2004). Goods and absolutes. In H. J. Gensler, E. W. Spurgin, & J. Swindal (Eds.), *Ethics: Contemporary readings* (pp. 222–228). Routledge.

Finnis, J. M. (2011). The basic forms of human good. In *Natural law and natural rights* (2nd ed., pp. 85–90). Oxford University Press.

Fletcher, G. (2017). Objective list theories. In G. Fletcher (Ed.), *The Routledge handbook of philosophy of well-being* (pp. 148–160). Routledge.

Frankfurt, H. G. (1971). Freedom of the will and the concept of a person. *The Journal of Philosophy, 68*(1), 5–20. https://doi.org/10.2307/2024717

Frederick, S., & Loewenstein, G. (1999). Hedonic adaptation. In D. Kahneman, E. Diener, & N. Schwarz (Eds.), *Well-being: The foundations of hedonic psychology* (pp. 302–329). Russell Sage Foundation.

Hall, A., & Tiberius, V. (2017). Well-being and subjective dependence. In G. Fletcher (Ed.), *The Routledge handbook of philosophy of well-being* (pp. 175–186). Routledge.

Heathwood, C. (2017). Desire-fulfillment theory. In G. Fletcher (Ed.), *The Routledge handbook of philosophy of well-being* (pp. 135–147). Routledge.

Hildebrand-Chupp, R. (2020). More than 'canaries in the gender coal mine': A transfeminist approach to research on detransition. *The Sociological Review, 68*(4), 800–816. https://doi.org/10.1177/0038026120934694

Hurka, T. (2009). The concept of human nature. In *Perfectionism* (pp. 9–18). Oxford University Press.

Kaltiala, R., Bergman, H., Carmichael, P., De Graaf, N. M., Egebjerg Rischel, K., Frisén, L., Schorkopf, M., Suomalainen, L., & Waehre, A. (2020). Time trends in referrals to child and adolescent gender identity services: A study in four Nordic countries and in the UK. *Nordic Journal of Psychiatry, 74*(1), 40–44. https://doi.org/10.1080/08039488.2019.1667429

Kruglanski, A. (1996). Goals as knowledge structures. In P. M. Gollwitzer & J. A. Bargh (Eds.), *The psychology of action: Linking cognition and motivation to behavior* (pp. 599–618). Guilford Press.

Littman, L. (2021). Individuals treated for gender dysphoria with medical and/or surgical transition who subsequently detransitioned: A survey of 100 detransitioners. *Archives of Sexual Behavior, 50*(8), 3353–3369. https://doi.org/10.1007/s10508-021-02163-w

Lucas, R. E. (2007). Adaptation and the set-point model of subjective well-being: Does happiness change after major life events? *Current Directions in Psychological Science, 16*(2), 75–79. https://doi.org/10.1111/j.1467-8721.2007.00479.x

Mill, J. S. (2009). *Utilitarianism*. The Floating Press.

Murphy, M. C. (2001). *Natural law and practical rationality* (1st ed.). Cambridge University Press.

Parfit, D. (1986). *Reasons and persons* (1st ed.). Oxford University Press. https://doi.org/10.1093/019824908X.001.0001

Railton, P. (1986). Facts and values. *Philosophical Topics, 14*(2), 5–31.

Rice, C. M. (2013). Defending the objective list theory of well-being. *Ratio, 26*(2), 196–211. https://doi.org/10.1111/rati.12007

Richards, C., & Doyle, J. (2019). Detransition rates in a large national gender identity clinic in the UK. *Counselling Psychology Review, 34*(1), 60–66. https://doi.org/10.53841/bpscpr.2019.34.1.60

Riis, J., Loewenstein, G., Baron, J., Jepson, C., Fagerlin, A., & Ubel, P. A. (2005). Ignorance of hedonic adaptation to hemodialysis: A study using ecological momentary assessment. *Journal of Experimental Psychology: General, 134*(1), 3–9. https://doi.org/10.1037/0096-3445.134.1.3

Rosati, C. S. (1995). Persons, perspectives, and full information accounts of the good. *Ethics, 105*(2), 296–325.

Rosati, C. S. (2020). Welfare and rational fit. In R. Shafer-Landau (Ed.), *Oxford studies in metaethics* (Vol. 15, 1st ed., pp. 241–262). Oxford University Press. https://doi.org/10.1093/oso/9780198859512.001.0001

Ryff, C. D. (1989). Happiness is everything, or is it? Explorations on the meaning of psychological well-being. *Journal of Personality and Social Psychology, 57*(6), 1069–1081. https://doi.org/10.1037/0022-3514.57.6.1069

Schroeder, S. (2017). Health, disability and well-being. In G. Fletcher (Ed.), *The Routledge handbook of philosophy of well-being* (pp. 221–232). Routledge.

Sumner, L. W. (1999). The desire theory. In *Welfare, happiness, and ethics* (pp. 113–138). Oxford University Press. https://doi.org/10.1093/acprof:oso/9780198238782.001.0001

Tännsjö, T. (1998). *Hedonistic utilitarianism*. Edinburgh University Press. https://doi.org/10.1515/9781474473057

Tov, W. (2018). Well-being concepts and components. In E. Diener, S. Oishi, & L. Tay (Eds.), *Handbook of well-being* (pp. 1–15). DEF Publishers.

Ubel, P. A., Loewenstein, G., Schwarz, N., & Smith, D. (2005). Misimagining the unimaginable: The disability paradox and health care decision making. *Health Psychology, 24*(4S), S57–S62. https://doi.org/10.1037/0278-6133.24.4.S57

Vandenbussche, E. (2022). Detransition-related needs and support: A cross-sectional online survey. *Journal of Homosexuality, 69*(9), 1602–1620. https://doi.org/10.1080/00918369.2021.1919479

Wiepjes, C. M., Den Heijer, M., Bremmer, M. A., Nota, N. M., De Blok, C. J. M., Coumou, B. J. G., & Steensma, T. D. (2020). Trends in suicide death risk in transgender people: Results from the Amsterdam cohort of gender dysphoria study (1972–2017). *Acta Psychiatrica Scandinavica, 141*(6), 486–491. https://doi.org/10.1111/acps.13164

World Health Organization. (2020). *Basic documents: Forty-ninth edition (including amendments adopted up to 31 May 2019)*. https://apps.who.int/gb/bd/

Chapter 9
Closing Remarks: Voluntary Limb Removal and the Challenge to Patient Autonomy

9.1 Introduction

The propositions presented in this book are intricate. On one hand, there is an ongoing debate about personal autonomy and the capacity for making medical decisions. As discussed throughout the chapters, the concept of autonomy varies depending on the perspective taken. For this study, a particular interpretation of individual autonomy was selected, aligning with the definition proposed by Beauchamp and Childress. This view emphasises the importance of making informed and voluntary decisions, highlighting the principles of self-determination and self-governance.

> At a minimum, personal autonomy encompasses self-rule that is free from both controlling interference by others and limitations that prevent meaningful choice, such as inadequate understanding. The autonomous individual acts freely per a self-chosen plan, analogous to how an independent government manages its territories and sets its policies. (Beauchamp & Childress, 2013, p. 101)

The choice to adopt this specific definition stemmed from two key considerations. First, the profound influence of Beauchamp and Childress on the ethical discourse around autonomy is undeniable. Their work has significantly shaped the ethical framework within which autonomy is both understood and applied across various contexts (Entwistle et al., 2010; Lysaught, 2004; Slowther, 2007). Second, the apparent simplicity of this definition made it highly practical during the fieldwork phase. A clear and concise working definition was crucial to provide a common foundation for discussions and analysis.

On the other side, the captivating matter of Body Integrity Dysphoria and the profound turmoil it creates across various areas. Recognising Body Integrity Dysphoria (BID) in an official medical taxonomy, such as the one endorsed by the World Health Organization, prompts responses that extend beyond medical discourse, delving into the very essence of the human condition with almost transcendental implications. This condition provokes deep reflection on identity, personal

L. Loriga, *Body Integrity Dysphoria and the Ethical Dilemma of On-Demand Amputation*, https://doi.org/10.1007/978-3-031-82443-2_9

autonomy, the boundaries of medical intervention and the increasingly blurred line between what is considered elective and non-elective medical procedures.

When autonomy and BID are analysed together, stimulating results emerge. The extent to which an individual can dispose of one's body is highlighted, challenging concepts such as beneficence, non-maleficence and justice. These findings are intriguing and provocative, casting a shadow on elements which are generally taken for granted. Suppose an individual wishes to have a limb removed to avoid painful medical treatment; such a situation seems straightforward, as no opposition should hinder any rational individual from making an informed, uncoerced choice about one's medical future. The present research project, however, overturns such an assumption. Despite the stable affective state of the patient and one's satisfactory understanding of procedures and implications, amputation is still discarded in many instances as a possibility of care, preferring other routes which get dangerously close to therapeutic obstinacy with complete disregard for the patient's wishes. The trade-off between long-term health goals and immediate pain relief seems to be the physician's default interpretation for many similar instances, which highlights, even more, the intricacies of who should have the final word on what is in the patient's best interest. These nuances represent just a fraction of the queries that have arisen while pursuing this research project's aims and objectives. They emphasise the complexity of medical decision-making and exemplify the many issues encountered while navigating this intricate domain.

After retrieving and addressing the three core questions that drove the research, the present conclusive chapter will reflect, summarise the findings and frame them within the current body of knowledge on BID and personal autonomy in the medical decision-making context. This approach provides a transparent statement of how this project wishes to bring added value to the topic, expanding moreover the scientific corpus of fields such as bioethics and medical anthropology. Following this evaluation, the chapter will pivot towards a discussion on avenues that hold the potential to further enrich the discourse on BID and personal autonomy within the medical practice. These new research opportunities will be highlighted, offering a roadmap for future investigations in these areas. Finally, as a fitting conclusion to this chapter and the research journey, a subjective and reflective closing remark will encapsulate insights and observations from a strictly personal perspective.

9.2 Starting Propositions

Three questions were proposed at the beginning of the project: How is the biomedical principle of autonomy interpreted and assessed from the medical standpoint regarding the aetiology and treatment of elective medical bodily modifications? To what extent does a patient's autonomy conflict with the physician's obligation of beneficence and non-maleficence? How do physicians discriminate between legitimate and illegitimate requests of body-modifying intervention?

The first question aimed to establish a framework; in other words, its scope was to delineate boundaries to differentiate criteria for elective types of medical interventions. Such a framework was deemed necessary for two reasons: BID amputation often falls within the realm of electiveness, even though such an assumption is incorrect. Nevertheless, such an analysis served to create a comparison between BID and all those practices recognised as not being a medical necessity (although using such a definition is dangerous) to demonstrate that despite the wrong categorisation of BID as a fancy rather than a need, the refusal to amputate still does not hold critical scrutiny as many other surgeries recognised today as being elective, holds similar if not greater risks (see for examples the growing concerns around Brazilian Butt Lift Del Vecchio & Dayan, 2019; Pazmiño & Garcia, 2023). Additionally, by establishing criteria of autonomy interpretation and assessment for elective-type medical intervention, it was possible to better reflect, by comparison, on non-elective type of medical interventions, which ideally hold less strict criteria. Suppose such an assumption is true (which it is, as discussed so far). In that case, there are still fewer reasons to reject amputation than there would be if BID's amputation was recognised as an elective type of medical intervention. Such a gap in logical consistency was filled by answering this very first query.

The second question's scope was to implicitly tackle the idea of the patient's best interest by analysing what elements, in the patient's request (hence one's narrative) to undergo any medical intervention (elective and non-elective indiscriminately), are selected and interpreted by the physician as trigger factors for reflection on balance between deontological and subjective obligations of beneficence and non-maleficence. Such an aim was deemed necessary as it served to investigate how respondents define such concepts and how they apply them during daily medical practice. Similarly, by tackling the idea of the patient's best interest, it was possible to covertly investigate also the extent to which, and in what ways, physician authority clashes with the patient's autonomy. In other words, what are the elements the physician utilises to push back specific patients' requests? By touching the patient's best interest, the second query also served as a guide to investigate medical paternalism.

Question number three finally pulled together the threads of the previous two, providing further complements and empirical evidence. Such a question served to take a step back, moving from the particulars of question one and question two to an overall view. By doing so, it was possible to see how the elements previously identified, such as the patient's best interest, beneficence, non-maleficence, autonomy, paternalism, deontological stances and subjective inclinations, conveyed towards practical examples. Such a channelling of a multitude of concepts and ideas served as a ground for observation and analysis of how such elements interact with each other and are arranged in the physician's narrative and the results they produce. In other words, such a question was used to explore how the subjectivity and agency of the physician, with all that such vast concepts entail, navigate the intricate realm of deontological duties originating from the medical profession and subjective beliefs or inclinations originating from subject-specific life occurrences. Combining these two elements was deemed necessary to maintain the subjectivity of the respondents and not categorise them solely based on their respective medical professions.

9.3 Delineating Autonomy

Elective medical therapy encompasses interventions categorised as non-urgent that are not directly linked to the immediate survival of the patient. Such a term implies a certain degree of autonomous choice from the patient, as opposed to non-elective interventions where the patient often faces limited alternatives—typically the intervention or facing more severe health consequences, even potentially life-threatening ones such as severe illness or death. However, acknowledging that not all elective surgeries are non-life-saving is crucial. While not directly tied to immediate survival, some elective procedures can significantly enhance a patient's overall well-being. This creates a complex web of categorisation where the boundaries between electiveness and non-electiveness blur. In light of such issues, it may be advantageous to reformulate the definitions of elective and non-elective medical interventions. A non-elective intervention is any procedure that is life-saving in the short term; it is an intervention that, if not performed promptly, could lead to the patient's demise. Simultaneously, a non-elective intervention can encompass procedures that substantially improve the patient's well-being, with health being considered part of the broader well-being spectrum. Therefore, a non-elective intervention differentiates between mere survival and thriving. Conversely, an elective medical intervention can be seen as any procedure that is not directly linked to the immediate survival of the patient; it encompasses interventions that do not significantly enhance the patient's well-being. An elective intervention distinguishes between living and living slightly better than before. Such an analysis can be said to be setting the boundaries between elective and non-elective types of medical intervention.

Suppose now that such a distinction would not be as elaborate but rather would be a superficial reflection. Even in that case, according to all the markers identified to distinguish between autonomous agents and non-autonomous agents within the simplest definition of elective medical treatment, the BID would still pass the test of surgery acceptance. Such markers are a one-to-one correlation between motivations and expectations, impulsivity and emotional stability. These elements are gatekeepers in distinguishing between patients deemed capable of making medical decisions about their future and those who are not. During the discussion, two significant questions surfaced. Firstly, can a patient's post-surgery regret be predicted, particularly for irreversible conditions such as Body Integrity Dysphoria? The consensus was that despite thorough assessments of a patient's medical history and psychological state, absolute certainty regarding post-operative behaviour is unattainable. The goal, instead, is to minimise reasonable doubts and risks. Secondly, the debate centred around postponing amputation in anticipation of potential future medical therapies. Here, the morality of such a possibility was questioned. It was noted that understanding one's situation, recognising suffering and demonstrating motivation for self-improvement to enhance overall well-being play pivotal roles in justifying the procedure. BID's characteristics fly right through such markers, which makes BID's amputation an acceptable option. This, however, is still not the case, and here lies a problem of logical consistency. The only element that could defy such an accusation is another core concept that emerged during the research phase: body harmony. This latter appeared as critical for autonomy assessment, which compares to ideas such as beauty, health and nature: harmony is to beauty as disharmony is to ugliness, and beauty is to health as ugliness is to harm. It suggests that health is intrinsically related to nature, while harm is associated with artificial interventions. This assessment implies an inherent notion—a belief in a natural way of existing, characterised by having a functional and harmonious body. In other words, there is a common understanding that a state of well-being is closely tied to the idea of a body that functions in harmony with nature. Such an element is crucial and is not limited only to elective types of medical intervention but amply (if not completely) overflows also in all other medical practices taken into consideration during the research. The juxtaposition of these concepts underlines the complex interplay between societal perceptions of beauty, health and the human body's natural state. It reflects a broader consideration of how individuals and societies define and value the concept of a normal or harmonious body and how this understanding influences medical decision-making within the context of surgical procedures. Such elements, moreover, defy the logical flow identified with the previous markers, all of which are rooted in methodical analysis. Among all of the markers identified, the idea of body harmony seems to go beyond logical necessitation and rather be rooted deep down in cultural and personal preconceptions about the right way of life.

It is possible to offer the following explanation to answer how the biomedical principle of autonomy applies to elective body-modification procedures. The interpretation and assessment of the biomedical principle of autonomy within the context of elective bodily modifications can be seen as a hybrid concept situated at the intersection of being a characteristic of the agent, assessed through a series of markers (or loops in which the individual needs to pass through), and intrinsically tied at the same time to specific actions or requests. In this latter case, some instances hold specific autonomy characterisations that disqualify the autonomy of the individual per se, as it is for BID's amputation. The existence of such a twofold view of autonomy presents intriguing implications in terms of what is ordinary and what is extraordinary within elective medical procedures. Requesting amputation, for example, retrospectively casts a doubt on the individual's rationality, undermining previous assessments of autonomy despite the fact that amputation, at least in the BID case, does not fall under the umbrella of elective medical procedures. At the same time, however, SRS is seen as elective due to its non-life-saving nature. Contemporary medical understanding, however, recognises it as a valid and effective treatment for Gender Dysphoria, emphasising its significance for the well-being of transgender individuals despite the invasiveness of its approach.

9.4 Beneficence Versus Non-maleficence: Which Principle Prevails?

To understand the extent to which an autonomous individual can dispose of one's body within the medical paradigm, it is necessary to define first duties and obligations of the medical practitioner. Such elements are crystallised in beneficence and non-maleficence.

> The principle of non-maleficence obligates us to abstain from causing harm to others … A harm is a thwarting, defeating, or setting back of some party's interests, but a harmful action is not always wrong or unjustified. (Beauchamp & Childress, 2013, pp. 150–153)

Within this normative framework, non-maleficence primarily involves negative types of action, emphasising abstention or refraining from causing harm to others. Similarly, William Frankena (Frankena, 1973), as cited by Beauchamp and Childress (Beauchamp & Childress, 2013, p. 152), reiterates the same idea of the distinction between non-maleficence and beneficence by highlighting that while non-maleficence primarily involves abstention—negative action—beneficence focuses on promotion—positive requirements for action. These two elements were recognised as being the pivotal points around which the physician decides what is in the patient's best interest. Such an assessment, in turn, gravitates around another core concept, which is harm.

The concept of harm featured prominently among the responses of the research participants as the chief discriminating factor between accepting the patient's narrative and requests or blocking such a narrative and requests by leveraging on medical authority and an ideal better understanding of what really would benefit the patient. When debating the concept of harm, another noteworthy trend emerged, which was the predisposition towards negative action rather than positive one. Such a predisposition is crucial in considering the ethical implications of BID's amputation. If the intention is first and foremost to improve or restore the well-being of BID patients through amputation, then beneficence seems to be the preferred pathway as amputation is clearly the result of an initiative, in other words, a positive type of intervention. This, however, seems not to be the case, with the primary intention being, as emerged, to maintain the current status quo. Such a conclusion leads to the belief that there is an inherent preference for maintaining the current status of things, even though such a status might be harmful per se, rather than breaking such a balance through a positive type of intervention. The empirical data, therefore, highlighted two interpretations of the idea of harm. In the first interpretation, the meaning associated with harm is a concept caused by a positive action of the physician, such as performing an amputation for a case like BID. In the second interpretation, the meaning associated with harm existed before the physician's acknowledgement, and this concept can ideally be remediated through a positive action. However, harm in this sense is expressed through the negative action of maintaining the status quo, as the BID case exemplified. A positive action, therefore, can either cause harm or relieve from harm. Between these two senses, the former's instances were identified unanimously as having the most severe implications. Within such a view, the idea of non-maleficence is seen as a more rigid principle than the idea of beneficence. Such a view is valid for the BID case, but such logical connections did not maintain their mapping when posed under more careful scrutiny. An individual with melanoma can be saved through excisional surgery (positive action). The same individual could die if the melanoma is left untreated (negative action). In this case, it is evident how the short-term physical harm inflicted by the physician on the individual through excisional surgery is justifiable in light of the greater good coming from removing the melanoma. Such examples highlight how BID is once again an oddity compared to other ethically challenging instances, which recall the previously exposed idea of body harmony as a justification for the predominance of non-maleficence over beneficence.

The patient's autonomy, therefore, does conflict with the physician's duties and obligations, and such a clash is most evident when assessing the meaning of harm. Such conflict is not rooted, however, in an antithetical conception of what beneficence and non-maleficence are between physician and patient. Instead, such disagreement is rooted in the different paradigms used for their interpretation. This is dictated by the fact that medical practice confers authority to the evidence-based approach ruled by lab measurements and concepts of ideal biological functionality. This leads, in turn, to different interpretations of what is in the patient's best interest, exacerbating the debate between medical authority and personal preferences. This conflict does not necessarily arise from a paternalistic stance rooted in contempt towards the patient but rather from a genuine concern to protect patients from overly broad interpretations of well-being. Such an approach is not inherently wrong; instead, it serves as a necessary element to ensure that medical practice is conducted with careful consideration and ethical scrutiny.

9.5 Assessing the Legitimacy of Body-Modification Procedures

Having explored the interpretation and assessment of the biomedical principle of autonomy from a medical standpoint in the context of elective medical bodily modifications, as well as the conflicts between patient autonomy and the physician's obligations of beneficence and non-maleficence, it is possible to address the legitimacy of body-modifying interventions as it benefits the grey area in which on-demand amputations, either for BID or something else, are sitting today. The widespread reluctance to consider amputation as a possible treatment for BID can be traced back to the prevalence of the idea of non-maleficence over beneficence and a rooted conception of a harmonious body aligned with natural standards. The historical lack of understanding about the condition and scepticism regarding its authenticity must also be added. Many respondents believe that individuals experiencing BID might not consider amputation if presented with alternative courses of action. Such an assumption is likely valid; nevertheless, such reasoning is purely speculative. If it is true that if there was an alternative to amputation with the same benefit, it would likely be the preferred path, it is also true that such an alternative as of today does not exist. The introduction of one variable to such reasoning began to diverge opinions towards supporting amputation; such a variable was the patient's well-being (phrased as the quality of life during the interview phase) post-amputation. Despite some respondents remaining inflexible in rejecting amputation, the second variable introduced, which conferred the same attribute to BID as SRS has today, started raising the prospect of formally recognising amputation as legitimate care. Upon further analysis, however, it emerged that such a shift in opinions was rooted in an intention to act benevolently rather than a genuine understanding of patients' discomfort and respect for patient autonomy. In other words, what emerged was a tendency towards resignation when posed in front of a solid paradigm to justify amputation. Such resignation can be expressed as follows: Given the fact that, as of today, no therapy is available for BID except amputation, then amputation can be performed if proper legal and ethical safeguards are arranged. Such amputation, nevertheless, is to be considered a harmful practice going against the patient's best interest by negatively affecting one's natural bodily configuration and functionality. This exposed the limitations in the doctor-patient collaboration, with doctors' decisions exhibiting bias towards an almost unquestioned ownership of the patient's best interests. Such a view also highlights another topic which is worth mentioning once more. Requests rooted in psychological discomfort are summarily dismissed (exceptions emerged as expected from the Clinical Psychologists and the Psychiatrists). This observation suggests that the acceptance of surgical interventions (not only aesthetic or cosmetic type) to correct a functional body—or potentially functional with medical aid—is more likely to be accepted when the intervention does not significantly jeopardise the individual's original physiological functionality. A paradigm shift is necessary in the unique case of BID, in which amputation fundamentally alters the body's functionality in a negative sense. Rather than focusing solely on physiological functionality, the emphasis should expand to encompass the psychological aspect, moving from the concept of disability to that of impairment.

It is possible now to answer how medical professionals discriminate between legitimate and illegitimate requests for body-modifying intervention. Four elements are relevant to provide an answer: the differentiation between elective and non-elective is one element, the recognition of the autonomous status of the individual, the normative status of the practice sought and the recognition that psychology and physiology are not antithetical concepts. This distinction between elective and non-elective types of intervention provides a foundational framework for evaluating the legitimacy and extent of a request, which the recognition of the autonomous status of the individual must compound. This autonomy acknowledges that individuals can understand the risks and benefits of medical interventions and make choices based on their values and preferences, which can even mean prioritising psychological welfare over the physiological one. The normative status of the requested practice plays a significant role in evaluating legitimacy. This involves considering whether the practice is recognised and accepted within the medical and legal frameworks. In summary, the legitimacy of a request for body-modifying intervention hinges on a multifaceted evaluation that incorporates various elements; it would be incorrect to pinpoint what such elements are. Rather, as it has been done, it is necessary to highlight the gravitational nodes around which such elements are entangled. Medical professionals must balance the patient's autonomy and ethical responsibilities, ensuring that decisions are made in the best interests of the patient's overall health and well-being.

9.6 Why This Book Is Needed

The peculiarity of BID draws the attention of experts in different fields, mainly within the areas of Neuroscience and Clinical Psychology (Brugger et al., 2013; Kasten, 2023; Saetta et al., 2020; Sedda & Bottini, 2014) to uncover the behavioural and neural mechanism that leads a rational, physiologically healthy individual to attempt, with any means necessary, to have one's bodily metrics and configuration impaired. Such approaches, although enlightening to progress the knowledge of behavioural and neural pathways towards a better understanding of human existence in general and BID itself, lack to some extent the acknowledgement of the individual's subjectivity and one's framing within the broader sociocultural context of reference. In short, the BID individual is often seen as a specimen to be studied rather than an individual to be comprehended.

> The specimen body is an object of medical attention created by the particular path medical institutions followed toward modernisation, characterised by the rise of interrelated specialisations and the development and eventual reliance on sophisticated imaging technologies. (Jones, 2011, p. 74)

The social and ethical declinations of the phenomenon seem to have been left to the background as if they are not as crucial as the physiology of the disorder (although not entirely forgotten, for example, Bayne & Levy, 2005; Brugger & Lenggenhager, 2014; Gibson, 2023). The BID phenomenon is also fertile for an overall assessment of individual autonomy and the extent to which autonomous agents can dispose of one's body within the contemporary medical framework. Similarly, the interpretation and assessment of such autonomy, in other words, how it is understood and interpreted in daily medical practice, is of importance as the former cannot be contextualised and analysed without this latter element. The significance of such analysis was long overdue.

One of the crucial elements emerged during investigation and analysis is how such boundaries of autonomy understanding and assessment, although ideally established through a collaborative effort between medical professionals and patients, are instead, in most cases, predominantly dictated in a top-down fashion similar to the idea proposed by Hacking (2006, 2007) in which the individual is processed, almost like a product, through the medical assembly line, and labelled with specific nomenclatures bearing all the weight of preestablished and incontestable medical guidelines. What was noted throughout the project is also how the principles of medical deontology, at times, take a back seat, leaving room for subjective perspectives of medical practitioners, who often rely on consensus-based ideologies—rooted in social and cultural norms—to distinguish between legitimate and illegitimate forms of medical intervention. In such a dynamic, patients often find themselves in a passive and voiceless role, with minimal scope for initiative. Requests by patients who venture beyond these predetermined social norms often face stigmatisation, as evident in cases involving requests for amputations either for BID or more standard occurrences. This seemingly paradoxical situation challenges logical reasoning. It becomes evident that the medical rationale that guides the acceptance or rejection of specific requests holds solid ground up to a certain point, beyond which it gives way to categorical judgements that lack the same solid foundation. By bringing attention to this lack of logical consistency, this book and the project from which it originates aimed at a more critical evaluation of medical decision-making within the context of BID and the broader realm of on-demand amputations. Furthermore, by highlighting the peculiarity of medical reasoning in such practices, the research allowed for a comparative analysis between BID and that of Gender Confirmation Surgeries such as SRS and similar interventions. This comparative analysis was undertaken to stimulate discussions about self-amputation and other forms of body modification. Understanding these moral and ethical aspects can give lawmakers and the justice system valuable insights, enabling them to make informed decisions. A comprehensive study encompassing ethical and social dimensions offers an interdisciplinary perspective in a landscape where contemporary literature often fixates on BID's neurological and psychological aspects. Such findings can be used as the foundation for developing guidelines and public policies about BID, ensuring that decisions concerning treatment and interventions are rooted in robust ethical and social frameworks.

9.7 Limitations and Future Endeavours

It is essential now to acknowledge and address some of this project's limitations, both technical and practical. Technical limitations can be categorised into three distinct groups, each influencing the study in various ways.

The initial set of limitations pertains to the geographic scope of the research. Both the qualitative and quantitative components of the study were conducted exclusively within the national borders of the Czech Republic. This geographical constraint undeniably restricts the applicability of the findings to other regions. This limitation is not solely due to potential variations in medical practices and ethical perspectives among different healthcare systems, although such differences could indeed play a significant role. Instead, it primarily stems from the profound impact of social and cultural practices and beliefs on healthcare approaches. As evident from the results presented in the preceding sections, there is a marked bias rooted in cultural elements that might very well be specific to the sociocultural context of the Czech Republic. These sociocultural factors may vary considerably from one country to another or even from one healthcare system to another. The present research project does not account for such variability, highlighting the challenges associated with extrapolating and applying the results universally across borders. This observation stresses the need to expand the scope of research of this kind across Europe. Ideally, future research should involve sampling respondents from multiple countries, thus facilitating the development of a more comprehensive and unified approach to the concept of personal autonomy and its relationship to BID. This broader, cross-cultural perspective would contribute significantly to a more nuanced understanding of these complex issues and their implications for healthcare practices on an international scale.

In addition to the technical limitations, the practical limitations of the study must also be acknowledged. Primary among these limitations is the scepticism and reluctance of medical personnel to openly discuss specific sensitive topics, which posed a significant challenge throughout the research process. This inherent scepticism may have influenced the honesty and depth of responses, potentially leading to an underrepresentation of particular perspectives within the study. The reluctance of medical professionals to engage in open discussions about sensitive topics, especially with layperson—in other words, not a medical doctor—could have resulted in self-censorship or guarded responses, hindering the research's ability to fully explore the nuanced aspects of personal autonomy and BID within the medical context. This practical limitation highlights the complexities of conducting research in healthcare settings where professionals may be apprehensive about discussing ethically and emotionally charged subjects. Strategies to mitigate such reluctance and scepticism among medical personnel should be considered in future research efforts to ensure a more comprehensive understanding of the topic.

The research project encountered additional challenges stemming from the COVID-19 pandemic, which occurred between 2021 and 2022. This global health crisis disrupted various aspects of the research, particularly in terms of engagement with hospitals and private clinics. Several factors contributed to these disruptions, including security concerns related to the pandemic and the increased burdens placed on healthcare professionals during this period. As a consequence of the pandemic, access to specific medical settings and personnel may have been limited, potentially impacting the diversity of data sources. Restrictions to in-person interactions, changes in healthcare priorities and the redirection of resources to combat the pandemic may have affected the willingness and availability of medical professionals to participate in the study. These challenges may have led to a narrower scope of perspectives and experiences represented in the research.

It is important to address another limitation: The lack of comprehensive official data concerning the status of BID individuals who managed to obtain amputation. Such a position can be used to support the claim that amputation is not the best way forward for BID individuals and that the analysis presented within the present research lacks a fundamental element. The premises of such a claim are true; the conclusion, however, is incorrect. As mentioned throughout the project, the scope of the present work was not to advocate for BID amputation; instead, it is to test the boundaries of individual autonomy within the healthcare setting by using BID as an exemplary case. It is necessary, however, to address such potential concerns for transparency's sake. The literature at disposal today on the BID phenomenon testifies to the scepticism of such individuals to come out due to fear of repercussion, which results in a lack of post-operative data. In the few examples of the direct testimony of such individuals (or allegedly so), such reticence is a reoccurring element (for example, Furth & Smith, 2002; Stirn et al., 2009). It is also necessary to point out that there might indeed be other approaches, as of today, than amputation. For example, a recent study had some positive results with Augmented Reality (Turbyne et al., 2021); nevertheless, the promising few are still in their early phase, hence casting judgement on their efficacy is highly premature.

To enrich the overall discourse on these topics, several relevant directions for future research can be suggested. One significant avenue for future research might involve expanding the pool of respondents to include individuals who have been formally diagnosed with BID. Investigating individuals who recognise themselves as having BID but have not received a formal diagnosis or have faced challenges in obtaining a diagnosis could be another valuable research area. This research can potentially shed light on the clinical criteria and barriers surrounding BID diagnosis. By understanding the experiences of undiagnosed individuals, researchers could contribute to the refinement of diagnostic processes and identify systemic barriers that impede access to appropriate care. Research should also delve into how individuals with BID, both diagnosed and undiagnosed, perceive their roles in contemporary society. This exploration could reveal the social and psychological challenges they face, as well as their coping mechanisms and strategies for acceptance and inclusion.

Another essential element might be investigating the underground community of individuals who have chosen self-made solutions over medical treatment. This inquiry can potentially provide insights into the extent of stigmatisation, the reasons behind their decisions and the potential risks associated with DIY interventions. Understanding this aspect of the BID phenomenon can guide the development of harm reduction strategies and inform public health initiatives.

A comprehensive analysis of amputation requests, extending beyond BID cases, holds the potential to yield valuable insights into the motivations and experiences of individuals seeking such procedures for various reasons. Investigating their interactions with medical professionals and examining the broader sociocultural context surrounding their requests can contribute significantly to the development of medical ethics guidelines and decision-making protocols.

These suggested research directions could offer numerous opportunities to advance the understanding of BID, personal autonomy and medical ethics.

9.8 Final Remarks

When the present research started, the COVID-19 pandemic was afoot, which seriously hindered the premises upon which the aims and objective of research were founded, namely empirical research in the form of fieldwork, as in the most traditional fashion for any anthropological work. The restrictions and safety measures implemented to mitigate the spread of the virus had limited, right from the start, physical access to research sites and communities. These measures encompassed travel restrictions, social distancing and lockdowns, which made it impractical or even impossible in certain cases to engage in the customary immersive and in-person fieldwork methods integral to anthropological research. These challenges included inherent limitations in establishing deep relationships and collecting first-hand observations, a hallmark of the Malinowskian participant observation method that has long been central to anthropological research. Consequently, the research process had to pivot (at least in the initial phase) towards alternative methods and strategies that could still provide valuable insights within the constraints of the pandemic's impact on traditional anthropological fieldwork. The involvement of private clinics and medical associations played a pivotal role in this research endeavour. Equally indispensable was the collaboration with several informants who served as intermediaries between the researcher and these institutions. These informants, who possessed a deep understanding of the customary modes of engagement, proved invaluable in navigating the initial scepticism exhibited by the respondents. Through their mediation, I was able to establish a bridge of trust and communication, facilitating access to critical insights and data that would have otherwise been challenging to obtain. This collaborative effort, built on the expertise of these informants, not only overcame initial barriers but also enriched the research by offering a culturally nuanced perspective on the subject matter under investigation.

It would be inaccurate to attribute the challenges encountered during the research solely to technical aspects. Many of the difficulties were closely intertwined with the complexities of conducting research in an environment where I needed to gain proficiency in the intricacies of social and cultural nuances. These barriers presented formidable obstacles, impacting the ability to effectively communicate, build rapport and navigate the finesses of the research context. Consequently, they significantly shaped the research process and the interpretation of findings, highlighting the essential role of cultural competence in the success of anthropological research in unfamiliar settings. Paradoxically, these challenges emerged as integral drivers of the research process. Over time, these barriers transformed from obstacles into catalysts, playing a pivotal role in shaping the direction of the research. These challenges were essential in fostering a profound reflection on the research objectives and the practical methods required to investigate BID and personal autonomy. They compelled me to adapt, develop and apply new skills and methodologies, thus evolving from abstract propositions to concrete actionable strategies.

Finally, a closing step back must be made to reflect on what BID signifies beyond what has been presented so far. BID can be seen as the final product dripping out the bottom part of a multidimensional funnel which mixes within itself three separate elements of human existence. Such elements are the three bodies possessed by every individual, at least from an Anthropological perspective (Scheper-Hughes & Lock, 1987). At the core of this framework lies the individual body, as experienced phenomenologically; in other words, the lived, subjective sense of Self in relation to one's body. The second layer is the social body, which is a symbol inherently tied to an idea, such as nature, culture or society. Within the medical paradigm, for example, as emerged throughout the research, a whole and functional body symbolises health, while its opposite represents sickness (Huggins, 2016). The act of elective amputation, in the BID case, represents a stark departure from traditional norms and values associated with the body. This radical shift can be viewed as a symbolic rejection of conventionality and a willingness to embrace individualism, self-expression and the rejection of established traditions. The decision to voluntarily alter one's body through amputation challenges deeply ingrained cultural and societal norms about the sanctity of the human form. In this sense, it could be interpreted as a rebellion against the constraints of Modernism, which often emphasises conformity and rationality. Instead, on-demand amputation becomes an embodiment of postmodernism's rejection of grand narratives and its embrace of fluidity, fragmentation and a multiplicity of perspectives. It is at this point that the third layer asserts its relevance, that is, the body politic, which highlights the control of bodies—both individual and collective—in domains such as reproduction, sexuality, work, leisure and health. BID seems to take these three layers of bodily existence and boil them down into an empirical manifestation, that is, the amputation request. Such a request forces reflection on certain elements that perhaps were meant to be left hidden or disguised in order to be effective as they exemplify dynamics of daily structural violence that go unnoticed. By bringing the issue to the forefront, BID shines a spotlight on the hidden and often uncomfortable tensions between individual autonomy and societal norms, between personal identity and collective expectations. Perhaps that is why BID is so provocative as an idea and even more so as an officially recognised medical condition: because it challenges all-at-once power dynamics that were designed to be left untouched and hidden from attention and critical scrutiny. It compels to engage in conversations about autonomy, identity and the boundaries of acceptability, challenging the status quo and urging us to consider the profound implications of these dynamics on individuals and society as a whole.

References

Bayne, T., & Levy, N. (2005). Amputees by choice: Body integrity identity disorder and the ethics of amputation. *Journal of Applied Philosophy, 22*(1), 75–86. https://doi.org/10.1111/j.1468-5930.2005.00293.x

Beauchamp, T. L., & Childress, J. F. (2013). *Principles of biomedical ethics* (7th ed.). Oxford University Press.

Brugger, P., & Lenggenhager, B. (2014). The bodily self and its disorders: Neurological, psychological and social aspects. *Current Opinion in Neurology, 27*(6), 644–652. https://doi.org/10.1097/WCO.0000000000000151

Brugger, P., Lenggenhager, B., & Giummarra, M. J. (2013). Xenomelia: A social neuroscience view of altered bodily self-consciousness. *Frontiers in Psychology, 4*. https://doi.org/10.3389/fpsyg.2013.00204

Del Vecchio, D., & Dayan, E. (2019). To ban or not to ban "Brazilian butt lifts"? Plastic surgery, responsibility, and the public health. *Plastic & Reconstructive Surgery, 144*(5), 952e–953e. https://doi.org/10.1097/PRS.0000000000006189

Entwistle, V. A., Carter, S. M., Cribb, A., & McCaffery, K. (2010). Supporting patient autonomy: The importance of clinician-patient relationships. *Journal of General Internal Medicine, 25*(7), 741–745. https://doi.org/10.1007/s11606-010-1292-2

Frankena, W. K. (1973). *Ethics* (2nd ed.). Prentice-Hall.

Furth, G., & Smith, R. (2002). *Amputee identity disorder: Information, questions, answers and recommendations about self-demand amputation*. Authorhouse.

Gibson, R. B. (2023). Body integrity dysphoria and medical necessity: Amputation as a step towards health. *Clinical Ethics, 18*(3), 321–329. https://doi.org/10.1177/14777509231160398

Hacking, I. (2006). Making up people. *London Review of Books 28*(16)

Hacking, I. (2007). Kinds of people: Moving targets: British Academy lecture. In P. J. Marshall (Ed.), *Proceedings of the British Academy, volume 151, 2006 lectures* (1st ed.). British Academy). https://doi.org/10.5871/bacad/9780197264249.001.0001

Huggins, R. (2016). The addict's body: Embodiment, drug use and representation. In P. Vannini & D. Waskul (Eds.), *Body/embodiment. Symbolic interaction and the sociology of the body* (1st ed., pp. 165–180). Routledge. https://doi.org/10.4324/9781315569635

Jones, N. L. (2011). Embodied ethics: From the body as specimen and spectacle to the body as patient. In *A companion to the anthropology of the body and embodiment* (pp. 72–85). Wiley-Blackwell.

Kasten, E. (2023). The girl who wants to get rid of her left leg – Body identity dysphoria. *Healthcare, 11*(13), 1901. https://doi.org/10.3390/healthcare11131901

Lysaught, M. (2004). Respect: Or, how respect for persons became respect for autonomy. *The Journal of Medicine and Philosophy, 29*(6), 665–680. https://doi.org/10.1080/03605310490883028

Pazmiño, P., & Garcia, O. (2023). Brazilian butt lift–associated mortality: The South Florida experience. *Aesthetic Surgery Journal, 43*(2), 162–178. https://doi.org/10.1093/asj/sjac224

Saetta, G., Hänggi, J., Gandola, M., Zapparoli, L., Salvato, G., Berlingeri, M., Sberna, M., Paulesu, E., Bottini, G., & Brugger, P. (2020). Neural correlates of body integrity dysphoria. *Current Biology, 30*(11), 2191–2195.e3. https://doi.org/10.1016/j.cub.2020.04.001

Scheper-Hughes, N., & Lock, M. M. (1987). The mindful body: A prolegomenon to future work in medical anthropology. *Medical Anthropology Quarterly, 1*(1), 6–41. https://doi.org/10.1525/maq.1987.1.1.02a00020

Sedda, A., & Bottini, G. (2014). Apotemnophilia, body integrity identity disorder or xenomelia? Psychiatric and neurologic etiologies face each other. *Neuropsychiatric Disease and Treatment, 10*, 1255–1265. https://doi.org/10.2147/NDT.S53385

Slowther, A.-M. (2007). The concept of autonomy and its interpretation in health care. *Clinical Ethics, 2*(4), 173–175. https://doi.org/10.1258/147775007783560166

Stirn, A., Thiel, A., & Oddo, S. (Eds.). (2009). *Body integrity identity disorder: Psychological, neurobiological, ethical and legal aspects*. Pabst.

Turbyne, C., Koning, P. D., Zantvoord, J., & Denys, D. (2021). Body integrity identity disorder using augmented reality: A symptom reduction study. *BMJ Case Reports, 14*(1), e238554. https://doi.org/10.1136/bcr-2020-238554

Index

A
Alternate possibilities, 30, 59
Alternative treatments, 59, 65, 74, 76, 77
Autonomy, 2, 3, 8, 25–30, 38, 40, 41, 43, 45, 46, 50, 51, 63–66, 71, 72, 77, 95, 105, 109–122

B
Beneficence, 3, 8, 25–29, 110, 111, 114–115
Bioethics, 2, 7, 25, 28, 82, 110
Blame, 60, 62–66, 69–78, 91
Blameworthiness, 69–71, 74, 76, 77
Body harmony, 8, 39, 45, 46, 97, 113, 115
Body images, 7, 12–17, 43
Body modification, 12, 30, 31, 37, 118
Body schema, 7, 13, 15–17

C
Cosmetic surgeries, 8, 12, 14, 16, 27, 31, 32, 37–44, 46, 51
Culture, 7, 12, 16, 17, 43, 60, 61, 122

D
Desire-fulfilment, 8, 83, 89–96, 103–105
Doing otherwise, 76, 77

E
Embodiment, 6, 11–17, 122
Ethics, 8, 25–33
Expectations, 37, 39–42, 44–46, 50, 55, 62, 74, 86, 87, 89, 92, 95, 104, 105, 112, 122

F
Future research, 119, 120

G
Gastric banding, 8, 49, 54–55, 72
Gender, 1–7, 12–14, 25, 27, 30, 31, 33, 39, 92–94, 113, 118
Gender affirmation, 8

H
Health seeking behaviour, 8
Hedonism, 8, 83–90, 103, 105

I
ICD 11, 1, 11, 13, 31, 70
Identities, 1–7, 12–14, 27, 92, 93, 109, 122
Informed consent, 2, 29, 30, 39, 42, 52

J
Justice, 8, 25, 26, 31–33, 110, 118

L
Legitimacy, 8, 37–46, 64, 73, 115–117

L. Loriga, *Body Integrity Dysphoria and the Ethical Dilemma of On-Demand Amputation*, https://doi.org/10.1007/978-3-031-82443-2

M
Medical decision-making, 14, 27, 29, 40, 42, 43, 46, 50–52, 70, 81, 91, 110, 113, 118
Medical ethics, 26, 39, 120
Medical scepticism, 12, 62, 69–78, 119
Moral actions, 59–61, 66, 69
Moral agency, 8, 61–62, 66
Moral cognitions, 8, 60, 61, 66, 78
Moral reasoning, 8, 60, 61
Moral responsibility, 8, 59–67, 70–78, 81, 91
Motivations, 39–42, 44–46, 50, 53, 55, 56, 61, 69, 73, 76, 92, 112, 120

N
Narrative identity, 5–6
Neurology, 1, 8, 11–13, 15, 17, 118
Non-maleficence, 2, 3, 8, 25–28, 110, 111, 114–115
Normativity, 8, 37–46

O
Objective list theories, 8, 83, 96–103, 105
On-demand amputation, 3, 8, 59, 66, 69, 81, 90, 115, 118, 122

P
Patient competence, 8
Patient intentions, 64, 78
Patient rationality, 42, 81
Patient regret, 8
Patient rights, 2
Plastic surgery, 12, 14, 17, 27, 30
Postmodernity, 12, 16, 17, 43
Post-operative behaviour, 8, 112
Praise, 60, 62, 63, 66, 77
Pre-operative assessment, 49–56
Psychology, 43, 103, 116, 117

R
Reactive attitudes, 62, 96
Reconstructive surgery, 51
Risk assessment, 8, 49–56

S
Self, 4, 5, 7, 12–14, 16, 17, 31, 94, 105, 122
Self-amputation, 32, 118
Sleeve gastrectomy, 49, 54–55
Societal norms, 1, 38, 46, 60, 122
Societies, 12, 16–17, 26, 86, 113, 120, 122
Surgical ethics, 25–33
Surgical interventions, 14, 31, 33, 40, 92, 93, 116
Symptomatology, 5–6, 86

W
Well-being, 8, 25, 37, 81, 112

The manufacturer's authorised representative in the EU is Springer Nature Customer Service Centre GmbH, Europaplatz 3, 69115 Heidelberg, Germany. If you have any concerns regarding our products, please contact ProductSafety@springernature.com

Printed and bound by CPI Group (UK) Ltd, Croydon, CR0 4YY
30/09/2026
02225513-0005